SHOW ME DON'T
LESSONS FROM AI IN

*To my father, whose undisguised glee in every little achievement
gave me the confidence to write this book. And to my mother,
whose dedication and determination gave me the skills
to achieve things worth writing about.*

Show Me Don't Tell Me

Lessons from AI in Defence

DARRELL J J JAYA-RATNAM

Howgate Publishing Limited

First published in 2025 by
Howgate Publishing Limited
Station House
50 North Street
Havant
Hampshire
PO9 1QU
Email: info@howgatepublishing.com
Web: www.howgatepublishing.com

British Library Cataloguing-in-Publication Data
A catalogue record for this book is available from the British Library

ISBN 978-1-912440-74-0 (pbk)
ISBN 978-1-912440-75-7 (hbk)
ISBN 978-1-912440-76-4 (ebk – ePUB)

The views expressed in this book are those of the author and do not necessarily reflect official policy or position.

CONTENTS

FIGURES AND TABLES

FOREWORD

The United Kingdom and its allies remain among the world's most capable Armed Forces. They are equipped with advanced platforms and weapon systems, and their people are well trained and highly motivated. However, in today's security environment, military advantage cannot be secured by traditional means alone. Potential adversaries are investing heavily in emerging technologies, including artificial intelligence (AI), and are moving quickly to incorporate them into doctrine and practice. The effective deployment of AI may prove to be the single most decisive capability in modern warfare.

AI is not an abstract concept; it is already changing how decisions are made, how intelligence is analysed, and how operations are conducted. At its core, AI performs tasks that until recently required human intelligence. Properly applied, it can support commanders, analysts and operators by enhancing their ability to act faster, with greater precision, and at scale. Yet within defence, experience of AI remains limited, and expectations often oscillate between unrealistic optimism and deep scepticism. Both extremes risk delay, over-regulation, and ultimately a failure to seize the initiative.

There are already examples where AI has delivered advantage on the battlefield. In Ukraine, AI-enabled analysis of satellite imagery and drone feeds have allowed commanders to identify and target hostile units at a pace that would have been impossible by manual means alone. This combination of rapid machine analysis with human judgement has given Ukrainian forces a crucial edge in time-sensitive operations. By contrast, trials of AI-enabled logistics planning within NATO have revealed how easily trust can be lost. When poorly explained outputs conflicted with commanders' intuition, users defaulted to manual methods, and the technology was set aside. The lesson is clear: AI succeeds when it complements military understanding and provides transparent, usable outputs; it fails when it is imposed without sufficient context, testing, or trust.

Getting the procurement of AI right is therefore essential. The UK's defence strategy depends on turning ambitious visions into operational reality. The creation of the Defence AI Centre is a positive step, but skills gaps, slow adoption and rigid procurement practices threaten progress. If acquisition is not agile, and if it does not prioritise the right partnerships and talent, the UK risks a damaging gap between its strategic ambition and its actual capabilities, undermining its aspiration to lead globally in the responsible and effective use of AI for defence.

Dr Darrell Jaya-Ratnam makes clear that AI must be judged not by promise but by utility. The simplest AI that achieves the effect is often the best solution. Defence must resist being seduced by unnecessary complexity and instead focus on practical applications that integrate into existing processes. Success requires a partnership: military personnel providing operational context; and technical teams delivering solutions that are explainable, reliable, and usable. Only then can AI outputs be trusted as the basis for action.

The stakes are high. In the military environment, there is no steady state. Inaction can be as damaging as error. History shows that shock and defeat often drive the most rapid learning. AI offers the opportunity to shorten those painful cycles, to learn faster and anticipate more effectively than an adversary can. But it must be introduced with humility as well as ambition, recognising both the potential and the limits of the technology.

This book arrives at the right time. It challenges extremes, clarifies misconceptions, and offers practical guidance to commanders and practitioners who must integrate AI into the realities of defence. Through four brilliant examples it also reminds us that AI is not an end in itself, but a means to enhance human contribution.

It is to Darrell's great credit that he has explained these complex issues with clarity and conviction. His determination to help defence personnel optimise their ability to integrate AI reflects a clear ambition: to deliver a winning solution through the partnership of human judgement and AI.

Air-Vice Marshal (Retired) Bruce Hedley MBE
Former Director of Joint Warfare, UK Ministry of Defence

PREFACE

Déjà vu and frustration. These are the two feelings that caused me to write this book. Over the 35 years that I have been involved in defence I have seen many technological and conceptual 'next big things'. There was, in no particular order, the internet, the revolution in military affairs, non-lethal weapons, network-centric warfare, network enabled capability, digitisation, autonomy and so on. Most of these led, eventually, to improvements in capability. But this did not happen smoothly as people and businesses jumped onto the latest bandwagon (with the best of motives, mostly) and the defence enterprise had to balance the desire for innovation with value for money, risk, short versus long-term operational advantage, and the needs of enduring versus contingent operations.

I see a similar thing happening with regard to artificial intelligence (the déjà vu) except that the noise and spin around AI is greater than I have seen before, thanks partly to the ease with which those with strong opinions, but no practical experience, can get a platform and audience. Having the military go through this hype-cycle may leave them at a disadvantage for some time, and that frustrates me given this is not new.

As a member of the UK Ministry of Defence AI Ethics Advisory Panel, I was very taken with the view held by many of the panel that the pace of development was so high that it was inappropriate to impose any strong rules or constraints. Instead they favoured providing principles and holding people responsible for thinking through how they would apply the principles in their particular case. This put me in mind of my experience at strategy firm McKinsey and Co where frameworks and Harvard-style case studies were used to get both strategy consultants and their clients to think through how they might address challenges and develop or implement strategy. I also used case studies extensively when lecturing on strategy at Birkbeck College, University of London. Birkbeck specialises in courses for mature and part-time students, and the feedback was that the case studies were a very efficient method of learning. So I set about trying to see if our experiences in AI were case study material.

Arguably, we benefitted a lot from the AI bandwagon. Out of 24 proposals submitted between 2019 and 2023, half were funded. These 12 projects led, directly or indirectly, to the development and implementation of the four AI applications that form the core of this book. All four are niche but, unlike the large-scale projects still in progress, they have been through the end-to-end cycle from idea to utilisation. This means that, whilst they do not hold lessons for every aspect of AI in defence, they can be used right now as case studies to help military officers be more resilient to spin and hype when setting requirements for AI and judging proposals for AI.

People learn better from being shown things than from being told things. I hope the experiences and lessons in this book can show the reality of defence AI to those rising up through the ranks of the military now.

Dr Darrell Jaya-Ratnam, PhD

ACKNOWLEDGEMENTS

This book summarises the lessons learned from nearly a decade of work. As such there are many people to thank. First and foremost are the people within the UK Ministry of Defence and its agencies, such as the Defence Science and Technology Laboratory, the Defence and Security Accelerator, which provided funding. Some of this came at the start and helped prove the concept, such as with the DUCHESS and MALFIE artificial intelligence applications. Some of it came after we had developed the technology and helped to prove its value in the military domain. All of it was crucial.

Alongside the funders are the many serving military personnel, from the UK and its allies, who brain-stormed with us, encouraged us, corrected us when we went off on any number of nerdy tangents, and helped demonstrate the value of the AI in a range of military contexts, scenarios and vignettes. Security and commercial confidentiality prevent me from naming names or detailing the capability that we delivered together. But you know who you are and what was collectively achieved.

The experiences and lessons that form the core of this book came from the wider DIEM team who developed the AI and managed the projects. Special thanks go to Sarah Vincent-Major who was the driving force behind DUCHESS, Josh Stapleton who provided key technical insights and support, and Yasmin Underwood who acted as our internal user and analyst in support of testing, application and ethics. Others who played a role in the various AI experiences and learning include Gabriel Alberici, Justice Azubuike, David Chitty, Edmund Day, Nathan Francis, Dennis Glover, Alex Howard, Alex Joseph, Jaymisha Panchal, Ania Pasierb, Matthew Stewart and Andrew Thomas.

I owe a debt of gratitude to my publisher, Kirstin Howgate, and to my editor, Madeline Koch, for their patience and guidance as I navigated the challenges of writing my first book. Not the least of these was finding my writing voice, getting the hang of the house style, and totally underestimating the time it would take! Finally, I could never have done it without the support of my wife, Reema, and my lovely dog, Rufus, whose loving distractions kept me sane.

ABBREVIATIONS

AASOPs	autonomous-agent standard operating procedures
AATTPs	autonomous-agent tactics, techniques and procedures
AEW	airborne early warning
AI	artificial intelligence
AIS	Automatic Identification System
AWO	air warfare officer
C2	command and control
CFP	Collision and Formation Pattern
CMS	combat management system
CNN	convolutional neural net
COA	course of action
CoBP	Code of Best Practice
DASA	Defence and Security Accelerator
DC	decisive condition
DLOD	Defence Lines of Development
DoD	Department of Defense (United States)
DRL	deep reinforcement learning
Dstl	Defence Science and Technology Laboratory
FACs	fast attack craft
FIACs	fast-inshore attack craft
FNR	false negative rate
FPR	false positive rate
HARM	High-speed Anti-Radiation Missile
HAT	human-autonomy teaming
HMT	human-machine teaming
IP	intellectual property
IRGCN	Iranian Revolutionary Guard Corps Navy
JAIC	Joint AI Centre (United States)
JALLC	Joint Analysis and Lessons Learned Centre
JDP	Joint Doctrine Publication

JSP	Joint Services Publication
LLM	large language model
KDE	kernel density estimation
MADM	Multi-Agent Dialogue module
MALFIE	Machine Learning Fuzzy-Logic Integration for Explainability
MECE	mutually exclusive and collectively exhaustive
ML	machine learning
MO	modus operandi
MOD	Ministry of Defence (United Kingdom)
MoE	measure of effectiveness
MoFE	measure of force effectiveness
MoO	measure of outcome
MoP	measure of performance
MoSP	measure of system performance
MWC	Maritime Warfare Centre
NATO	North Atlantic Treaty Organization
NDM	naturalistic decision making
NLP	natural language processing
NNs	neural nets
OODA	Observe Orientate Decide Act
PSO	particle swarm optimisation
R&D	research and development
RAS	robotic and autonomous systems
RMS	root mean squared
RPD	recognition-primed decision
RN	Royal Navy
SA	situational awareness
SE	supporting experience
SI	swarm intelligence
SME	subject matter expert
SOPs	standard operating procedures
SVM	support-vector model
TTPs	tactics, techniques and procedures
TREAD	Traffic Route Extraction and Anomaly Detection

1

Introduction

First, artificial intelligence (AI) was a mystery. Then a miracle. Now it's about money – investing it and making it. But beneath the gold rush lurks a familiar cycle: overpromise, overspend, underdeliver. The last transformational technology to fall victim to this was the internet and the dotcoms that were built from it. Things took off in 1993–1995, money poured in, and 2000–2002 saw the crash. This poses two questions for defence. Is AI falling into the overpromise, overspend and underdeliver cycle? And what can defence do to minimise the risks, or perhaps avoid the trap entirely?

The answer to the first question appears to be a definite 'yes' on the basis that opinions and interest about AI are widespread and expectations unrealistic. By way of a personal example, on a recent taxi ride I was regaled with insights about AI by the driver. It brought to mind the stories of Joseph Kennedy (father of US president John F Kennedy) and Bernard Baruch (a successful investor), who famously withdrew their money from the stock market days before the 1929 crash after their taxi drivers and shoe-shine boys started giving them investment tips. Their rationale was that the market had become 'too popular for its own good'[1] and was, therefore, irrational. It may be harsh to label individual and corporate opinions about AI as irrational, but it is not difficult to find opinions that are, at very least, misinformed with regard to AI in defence. An example is the misconception that the war in Ukraine is a proving ground for AI in defence due to the false assumption that drones are synonymous with AI. Some drones may use some AI, but the vast majority of drones are remote-controlled (such as the 'First Person View' drones common in videos from Ukraine) or have an advanced version of the type of sensor and terminal-guidance systems that have been used in missiles

---

1 John Rothchild, "When the shoeshine boys talk stocks: It was a great sell signal in 1929. So what are the shoeshine boys talking about now?" (*Fortune*, 15 April 1996). Retrieved from https://money.cnn.com/magazines/fortune/fortune_archive/1996/04/15/211503/.

for 60 years. AI companies such as Anduril and Helsing are developing AI-driven autonomous drone systems, but these have not completed the cycle from technology concept to use by the military at scale.

Drones, of course, are just one of many applications of AI in defence. Peruse the online list of projects funded by the UK's Defence and Security Accelerator since 2017 and you will find literally hundreds of AI ideas that were funded to the tune of between £50,000 and £3 million. There were probably hundreds more that could not be funded and were taken forward elsewhere. These ranged from the use of AI to enable automated weapon systems and platforms all the way back to the use of AI to manage supply chains and conduct exit interviews. For many, especially within the military, the most interesting question was what AI is the most useful and what will be game changing. Writers such as P.W. Singer, Paul Scharre and Kenneth Payne, amongst many others, have provided different answers to this strategic question based on their particular viewpoints and interest. However, the lack of any obvious AI 'killer app' or game-changing in the current slew of conflicts would suggest there is no quick or easy answer to that question. Instead, it is likely that AI will follow the path of previous technologies such as the microchip, communications, the internal combustion engine and gunpowder: it will have some impact everywhere, but what is most useful and will change the game will not be known until it is adopted and used at scale. Even with something as defence-specific as the tank, it took a generation to develop the correct mix of technology and doctrine for it to become a game changer.

Hence, the first requirement is to deliver successful AI projects into the hands of the military. Alas, on top of the complications caused by misconceptions and breadth of opportunities (or perhaps because of them), there is the problem of excessive expectations. As the 'Gartner hype-cycle' for AI so beautifully illustrates,[2] most AI-related technologies are at or approaching the 'peak of inflated expectations' and about to enter the 'trough of disillusionment'. This results from organisations throwing lots of money at a technology for fear of falling behind, spending that money unwisely, and then being disappointed by the results. This leads to a backlash that constrains the successful exploitation of the technology for a while. Eventually organisations go up the 'slope of enlightenment' about what

---

2 Afraz Jaffri, "Explore Beyond GenAI on the 2024 Hype Cycle for Artificial Intelligence", last modified 11 November 2024, https://www.gartner.com/en/articles/hype-cycle-for-artificial-intelligence.

the technology can do and how to use it, before they reach the 'plateau of productivity' where it is delivering actual benefit. In most sectors this cycle is no more than wasteful in money and time. In defence this could lead to projects being cancelled and finding oneself at a disadvantage compared to one's enemies – with all the risks to life and national interest that brings.

So, to the second question – what can defence do to mitigate or avoid the overpromise, overspend and underdeliver cycle? There are few, if any, end-to-end examples of large-scale AI use in defence which are sufficiently unclassified to learn lessons from. There are, however, many examples of small-scale and niche AI applications that have successfully gone through the end-to-end cycle of concept through to defence use. This book uses four such examples to provide some fundamental answers to the question of what defence (particularly military officers) can do to mitigate or avoid the overpromise, overspend and underdeliver cycle. Something this book does not do is try to advise about AI development approaches or techniques. As with all production processes, the companies involved will learn and improve over time as the technology changes.

There are three fundamentals aspects that military officers can and should drive, and which the four cases of successful AI implementation bring out. The first is the preparation of both the military and technical stakeholders in terms of their shared understanding of the use case for the AI. The second is the evaluation of the AI technology in the context of that use case including whether it is at the right level of complexity (not too much, not too little). The third aspect is the implementation of the application in a way that takes into account the broader system or environment.

Challenges

Imagine yourself in the shoes of a military officer in charge of developing the concept of, or setting the requirements for, a defence AI application. Perhaps you are such an officer or have aspirations to be one. Budget is not a problem for you. AI is the current 'big thing' and your command has thrown a lot of money at you. Naturally, you want to get the most for your money so you search out the views of those within your defence research and development (R&D) organisation. They wow you with the latest AI techniques[3] and update

---

3   As of 2025 this is generative AI and large language models, as popularised by ChatGPT. Prior to that it was generative adversarial networks, neural nets and deep reinforcement learning. Agentic AI systems are emerging as, potentially, the next big 'AI thing'.

you about the various innovation hubs, accelerators and AI centres through which the leading-edge applications can be accessed. Of course, 'leading edge' can often mean 'immature' in terms of robustness and reliability, but the vision of what is possible excites, and you buy into the potential benefits that the most advanced AI can bring. This may be at the expense of time and money, of course, but then military acquisition programmes are always long, and even the fastest will be greater than the small number of years you are likely to spend in the post anyway.

Then, being a diligent and intellectually curious sort, you also explore what the wider community of experts and commentators are saying. This opens your eyes to all sorts of concerns, from the risk of bias in the AI and the ethics related to using the AI for defence (which might relate to death and destruction) through to the apocalyptic visions of AI taking over the world. Would you be held responsible for any of these; might they come back to haunt your career? Is all this just the modern equivalent of 'FUDing' (where FUD = Fear, Uncertainty and Doubt, a technique used in the marketing and sales of technology in the past)? It may be that a relevant AI conference to explore these issues takes place, similar to the 2023 AI Safety Summit hosted by the UK Prime Minister, Rishi Sunak. But chances are that you will observe that the main speakers are the visionaries and leading tech-firm bosses whom the politicians want to be seen with (the equivalents of Elon Musk and Sam Altman) rather than individuals with first-hand experience of delivery. And you might also observe that despite the wide and balanced discussion of the issues that took place, the reporting focusses on what will get most coverage in the near-term rather than impacts over the long term.

At this point you might look for other military officers who have exploited AI before. Are there simpler technologies which can be used to solve the particular use case you have? Do all the risks and issues raised apply to your AI application? However, your advisors' responses are always 'I am not aware of something like this being done before'. Of course, you know that the absence of proof is not proof of absence, and that large organisations can be stove-piped to such an extent that people in one area may not always be aware of what has been done elsewhere. You may even suspect that any previous work may be being discounted in the belief (or desire) that the programmes being pushed by the current crop of experts are novel and innovative, and to avoid being constrained by previous work. However, you are where you are, and you need to progress your responsibility, so you decide not to delay further by continuing to search for what may have been done before.

Therefore, you take on the cost and complexity of leading-edge technology, and the risks of bias, ethics and unintended consequences. Meanwhile, in another part of the ministry, another officer is going through the same thing.

The years pass by and the programme reaches its conclusion. You may not still be there as the two-year posting cycle would have taken you elsewhere, but you might check up on the progress to see what impact you had. Or perhaps enough years have passed for you to return to the project but at a more senior level. What will you find?

The history of technology development in general (let alone in defence) suggests that what you will see is that the AI technologies used, which were at or approaching the peak of inflated expectations when you started, are now deep in the trough of disillusionment. As a result the project did not deliver the promised cost-benefit, the money to try again has gone or been severely cut back, and other technologies now attract all the attention. Or, by applying some of the lessons highlighted from past successful projects, you might see the AI you initiated or procured having an impact on capability and in operations, and being a reference case for AI that moved to the slope or enlightenment (maybe even into the plateau of productivity) more quickly than everyone else.

Digging the Trough

It is worth considering what could cause the trough of disillusionment for AI, particularly in defence. Fundamentally, war is a human endeavour undertaken with as many material and conceptual additions as the participants consider necessary and acceptable to overcome human limitations. Wars begin when some people decide to challenge other people's rights to govern, space, resources, life or worship. Wars end when one of these two groups decides that it lacks the skill or will to continue challenging or resisting. The part in between these human bookends is what warfare is about and, ever since the first primitive humans picked up a stone or stick to aid the fight, there has always been some element of 'artificiality' to it. At the most basic, functional, level all military 'technologies' – be they spears, slings, the spear thrower, bows and arrows through to rifles, artillery, war planes, radar, ballistic missiles and the atomic bomb – are artificial means of enabling one to fight faster, further, for longer and more safely.

Over the last 40 years, research into long-standing capabilities such as small arms, cannons, mortars, bombs, armour and aircraft has been joined by research into digital technologies such as Network Enabled Capability (related to the US concept of Network Centric Warfare) and decision tools to support logistics. In all cases, the research was about making either the platform or the weapon go faster, further, last longer or give a bigger bang, and demonstrating that you had achieved any of these was conceptually straightforward. That is not to say that all such technologies make it into service. Some research on various technologies has lasted decades and never resulted in anything being put into service. The reasons were almost always that the procurement process (from concept to requirements to development) took so long that either the enemy or the government changed and, as a result, it was no longer a priority, not cost effective or not affordable. Certain areas, however, benefited from good timing: the fall of the Berlin Wall, Gulf War I, the wars in the Balkans, and the long deployments of British troops in Afghanistan and Iraq created a sense of urgency that meant that some people saw several of the research projects they worked on being used in systems that ended up in the hands of troops. The Russian invasion of Ukraine seems to have created a similar sense of urgency to enhance military capability so AI would appear to fall into the latter category.

AI technology could be considered different to traditional military technology in that it seeks to achieve better speed, range, persistence and safety by directly improving the cognitive aspects of decisions previously made using natural or human intelligence, for example whether to use something, when to use it or how to use it. In terms of the decision cycle, also called the Boyd cycle or OODA loop (Observe, Orientate, Decide, Act), communication and sensor technologies have been artificially boosting the Observe stage, whilst chemical energy and computing power have been artificially transforming the Act stage (along with other technologies). The Orientate and Decide stages remained the province of human intelligence until communication, sensor and computing technologies, plus data and knowledge capture, all reached a sufficient level to combine and allow what we now call AI to transform the Orientate and Decide stages.

For a range of reasons – perhaps the widespread application of AI compared to the more niche applications of other technologies, maybe the greater ease of participation in AI for defence, possibly the potential for crossover from civil to military AI applications (and vice versa) and or because of the speed with which AI could be developed and used – there has been much greater and wider discourse about the military use of AI than for any other technology. Ironically, the increased profile of 'defence AI' in the public discourse might help dig a deeper trough of disillusionment than normal, and is driven by both those who are against AI's use in defence and those who are for it. These are the 'hyperati' whose business model is to benefit from exaggerating both the risks and opportunities of AI. On the one hand are the doomsayers who have tarred all potential AI (and AI usage) with risks that are only relevant to a subset of AI and their use cases. This leads to militaries spending more time and money proving the risks are not as high, or can be managed more easily, than the actual risks warrant. On the other hand are the AI devotees who, by focussing on the leading-edge AI (rather than the most easily exploitable AI), are also causing the military to spend more time and money, but this time on making immature and complex AI work well and safely.

What if the same approach was taken for other technologies? The most common cars in the United Kingdom are the Ford Fiesta and Volkswagen Golf, whilst the most popular vehicle in the United States is the Ford F-series pickup truck. What if, however, government traffic and transport strategy and implementation were based on the risks and potentials of a Bugatti Chiron? Yet this is, too often, how AI strategy is being driven – the two tribes of the hyperati talking up or talking down the most advanced AI, leading to policies based on that rather than the most common, relevant or useful AI.

The Technology Complex

A significant driver of both extremes is the focus on technology as being 'decisive' and 'war winning', and the hype around AI (whether alarmist 'we will lose control of it and it will destroy civilisation', or enthusiastic 'we will always win and never lose anybody') is just a part of this trend. The 1991 Iraq war has been, arguably, the poster child for this view, causing much ink to have been spilled on the 'Revolution in Military Affairs' amongst other things. Similarly, the repeated statements that, first, western anti-tank missiles would be a game changer in the Russo-Ukraine war, then Patriot missiles, then Leopard, Challenger and Abrams tanks, then ATACMS missiles, then F-16s, and so on.

But a deeper and wider look reveals that the decisiveness of technology alone is less clear. Stephen Biddle[4] makes the case for the importance of

---

4 Stephen Biddle, *Military Power – Explaining Victory and Defeat in Modern Battle* (Princeton University Press, 2006).

doctrine and tactics over technology, but here are a few simple examples, starting with Gulf War I. Aside from the technological superiority of the US and its allies over Iraq, the military build-up that took place prior to Desert Storm meant Iraqi forces were also outnumbered. In addition, Iraq was exhausted after the long war with Iran and the Iraqi military's experience was primarily in trench and urban warfare in the mountains and cities in the east, rather than mobile armoured warfare in the deserts of western Iraq. Finally, Iraq was surrounded by hostile states with no safe rear area.

By contrast, in World War Two, Nazi Germany was the first to deploy most of the technologies now considered part of modern warfare, from jet engines to guided missiles and ballistic missiles, but lost the war. In Vietnam and Afghanistan, the US and its allies were a generation, or more, advanced in technological terms than either the North Vietnamese or the Taliban, but suffered strategic defeat.

The comment received from serving and ex-military officers in regard to Vietnam and Afghanistan is often that 'we/they never lost a battle – it was the politicians who took the decision to withdraw'. This may be true but the implication that the relevance of technological edge in these cases cannot be judged misses the point, for three reasons. Firstly, winning (or not losing) a battle, or achieving any other campaign objective, does not always directly correlate into campaign victory. Napoleon won the battle of Borodino – his Grande Armée suffered fewer casualties than the Russians and it was the Russians who withdrew. However, he did not win 'well enough' for it to be a decisive victory, which led to campaign failure. Similarly, the Tet Offensive ended with a devastating loss in men and materiel for the Viet Cong but not quickly enough to avoid the impact on US public opinion.

Secondly, it depends on the scope of the metrics used to define a 'battle won' in the context of the objective of stopping or changing an adversary's behaviour. For instance, when it comes to casualties, the West often ignores the casualties suffered by its local allies. When they are added, one could make a case that in recent wars 'our side' lost more than 'their side'.[5] Finally, politicians are the strategic end of the military endeavour. If they made

---

5   In the 20 years of operations in Afghanistan, the North Atlantic Treaty Organization (NATO) suffered fewer than 8,000 troop and military contractor deaths but up to 42,000 'opposition fighters' were killed. However, to be added to this are nearly 59,000 Afghan soldiers and police, whom NATO were training, arming and fighting alongside, plus 38,000 civilians whom NATO were ostensibly protecting and supporting.

the wrong choices when deciding to embark on, measure the success of or withdraw from a military campaign, they did so based on inputs – good, bad or indifferent – from military officers and advisors. If technological superiority were reliably decisive or war-winning, then it should be able to deliver battles (and other objectives) that are not just won but won so well as to provide favourable inputs to the politicians.

The expectation that technology will prove decisive appears to be greatest amongst politicians and the media. Within the military there is much more awareness that military success, let alone a decisive victory, is a complex mix of a range of factors. This is evidenced by the importance given to the Defence Lines of Development (DLODs) in planning, described through the mnemonic 'TEPIDOILI' (Training, Equipment, People, Information, Doctrine, Organisation, Infrastructure, Logistics and Integration). A particular technology on its own can drive increased capability in several DLODs in isolation. But for real war-winning capability, all the DLODs have to work together well enough for the increased capability in a particular DLOD to be exploited and deliver benefit.

The unrealistic expectation that AI technology per se can be transformative drives both the fears of the doomsayers and the enthusiasm of the devotees. Ironically, these two opposite views lead to the same outcome – unrealistic expectations that soak up a lot of money but are not met as a result of additional complexity which, in turn, leads to disillusionment and delays in the next round of projects that can learn the lessons.

By contrast, understanding that (as with previous technologies) transformation and decisive impact will only come after a period of experience building, and integration of AI technology with the other DLODs can speed its ability to have an impact through encouraging people to start simple, learn how to work it and see some practical, tangible, early benefits. This need for humans to 'work' the AI points to the initial value that AI can add. Far from rapidly replacing people, and turning war into a series of automated processes, AI in the near term can enhance the contribution that individual humans can make to defence. To put it another way, enabling the utilisation of AI drives the utility of AI. Ultimately the iteration between utility and utilisation will drive the standardisation and democratisation of AI. How long that takes depends on how well defence is able to avoid the overpromise, overspend and underdeliver trap.

Inspiration and Insights

The inspiration for this book came from a personal experience whilst working on a joint UK/US special forces R&D project around 30 years ago. As the technical leader of a small research team within the Defence Evaluation and Research Agency of the UK's Ministry of Defence (MOD), I was seated at a table with a lieutenant colonel from the US Special Operations Command and explained how the techniques developed in my PhD (in ballistics) could be used to design a piece of ammunition that would meet their specialist requirements. After I had finished, he turned and said, "I'm from Idaho, and that's a 'show me don't tell me' kinda place". He then proceeded to lay down a series of challenges which he wanted me to show that my design could overcome.

That wonderfully practical, if cautious, approach to the exploitation of ballistics theory seems not to have an equivalent in defence AI, yet. At the programmatic level there are the stages of proof of concept, proof of value, prototype and minimum viable product used to develop the AI technology itself. At another level, there are the acceptance criteria that ensure the AI application meets the requirement. However, in between these two levels, there is no consistent set of 'in-the-context-of-a-military-user' challenges. Part of this may be because, unlike the ammunition example above, there is little first-hand experience amongst the military of what to focus on and check for with AI. It may also be that fields such as information technology and AI lend themselves less to realistic practical demonstration, or that their growth out of civilian use creates a 'lent authority' to whatever is said by 'experts'.[6]

Ideally this book would present lessons from many AI projects including large-scale AI implementations that have been game-changers in defence. For various reasons, some of which I have alluded to, this is not currently possible. AI is still relatively new and the breadth and depth to which AI could be applied to the defence enterprise means that there are many small or niche examples but no unclassified large-scale ones. Additionally, work on the large-scale implementation of AI in defence is being led by large companies which, for reasons of competitive advantage,

---

6 There is also the risk that the experts have no first-hand experience of delivery, just a lot of aggregated knowledge gathered from multiple sources applied without understanding the specifics of the military context.

security and protection of intellectual property (IP), have little interest in sharing their lessons learned. Instead, this book examines four small but diverse AI projects, each delivered by a small company that was willing to be open about their experiences.

When collating the information from these experiences of implementing AI into defence, several themes emerged which are likely to endure irrespective of the speed and extent of change in the AI sector over the next decade, at least. Firstly, they all address problems and challenges that the people working in the area know exist, but have chosen to ignore or put up with until the AI offered a solution. This helps address the risk of overpromising caused by losing sight of the tangible problem being solved or opportunity being gained with the AI. Secondly, the examples cover a range of different AI techniques, some leading edge but some relatively old. This illustrates that what makes something AI is not the age or complexity of the techniques used (contrary to the marketing efforts of some companies) but the functionality achieved, specifically fulfilling a task that normally requires human intelligence. This helps address the risk of overspending caused by overcomplicating the technology used. Thirdly, developing and implementing AI successfully is a journey of realisation and trust building for the user as much as the developer. A key part of this was helping users understand that the AI would help them do more or do better (within the constraints they are operating within) rather than replacing or automating their work. This helps address the risk of underdelivering through unstated assumptions and misaligned expectations.

These themes were combined with the three aspects discussed previously (preparation, evaluation and implementation), to determine ten key questions that military officers can ask to climb the 'utilisation staircase' of AI, as shown in Figure 1.1.

The key insight from the examples of AI implementation in this book are that the normal assumption – that one should first prove utility and only then would get utilisation – did not apply. Instead, addressing the questions in the staircase helped achieve a critical level of utilisation of the AI applications. This, in turn, drives the familiarity and understanding needed to identify how to get the greatest utility from the AI. The staircase can act as a guide for military officers involved in the development and implementation of AI in defence, with the diverse examples given illustrating the practicalities of doing so.

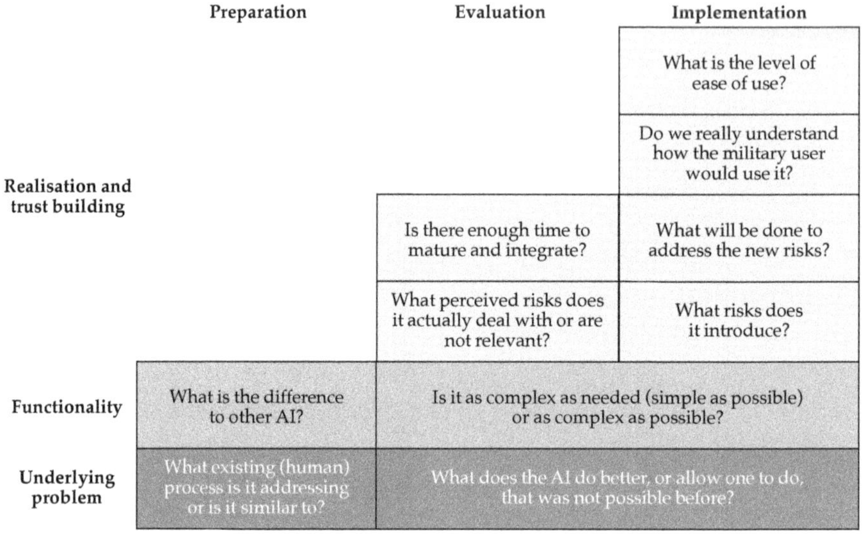

	Preparation	Evaluation	Implementation
			What is the level of ease of use?
			Do we really understand how the military user would use it?
Realisation and trust building		Is there enough time to mature and integrate?	What will be done to address the new risks?
		What perceived risks does it actually deal with or are not relevant?	What risks does it introduce?
Functionality	What is the difference to other AI?	Is it as complex as needed (simple as possible) or as complex as possible?	
Underlying problem	What existing (human) process is it addressing or is it similar to?	What does the AI do better, or allow one to do, that was not possible before?	

Figure 1.1: Utilisation staircase for AI

Structure

This book is divided into three parts. Part I can be considered to be theory, or possibly philosophy, starting with a chapter exploring what AI actually is. Part of the problem with the hype around AI (whether positive or negative) is that it is often defined in narrow, technical ways that suit the agendas of both extremes. There are, however, formal definitions that have been adopted by major organisations, such as the US Department of Defence (DoD) and UK MOD, which define AI in broader and more useful terms. These organisations have also gone on to lay the foundations of AI ethics and trust and have placed some of the greatest challenges to the issue of testing, which are the subjects of the next chapter. Grouped together are ethics and trust partly because each drives the other and partly because they often form the battleground amongst the tribes within the hyperati collective. Unfortunately, past and current work in AI ethics and trust is frequently disregarded (or assumed not to exist), leaving the poor military officer charged with moving AI forward in a particular area struggling to reinvent the wheel. Clearly, AI ethics and trust in AI are developing fields but it seems useful to highlight what the current policy is on these things. Also included are observations of what and how we trust human decision makers versus AI, which is useful when engaging with potential users of the AI.

Part II covers four first-hand accounts of the practical experiences of taking AI from concept to use with defence organisations. Of the AI applications covered in these four cases, one has been used by both military and commercial organisations, two by operational entities, and one by defence industry in the R&D of new capability. The first case, DUCHESS, concerns the use of AI to capture lessons learned, originally from military personnel who hads returned from deployment but subsequently in a range of military, commercial and even personal use cases. The technology preceded ChatGPT by many years and shows the value of using AI to leverage human intelligence. The second case, MALFIE, is an AI application that helps human operators prioritise between the outputs of multiple AIs – in this case for the monitoring of large areas of the world's sea lanes and oceans. Again, it shows that, however good the AI is, people need to take accountability and responsibility for the resulting actions taken, and how AI can help them. The third, Red's Shoes, relates to AI used to understand adversary commanders and decision-making bodies as part of operational and strategic planning. It illustrates how AI, which is often the subject of concerns over bias, can actually be used to help humans overcome their biases. The final case, DR SO, was originally developed as a 'threat agent', that is an example of AI that a potential enemy might deploy against us, but is subsequently being used to help develop AI algorithms for future systems and equipment. Again, this shows the potential of AI to augment current human-based processes at the 'back end' of the defence enterprise.

Each of these four case studies begins with a summary of the key takeaways in terms of how the steps of the utilisation staircase helped avoid the trough of disillusionment and eventually led to adoption. This is followed by an explanation of the context of the proposed application and its specific use case. These chapters then describe the inspiration for each application, the inception of the work, the technology, proof of concept, development and adoption. Lastly, there are reflections on what the experience of taking that particular AI application from concept into the hands of potential users might mean for AI in defence in general. Given the purpose of the book, and to avoid appearing as an extended marketing pitch, the description of the AI application or technology itself has been minimised. Naming clients and users, other than as part of their parent organisation, has also been avoided.

Again, this lessens the risk of coming across as marketing and ensures there are no issues with IP or security classification.

Part III brings together the lessons from the individual cases. This starts with a description of different approaches to testing AI and how this in turn drives ideas about verification and validation, assurance and qualification. This is important because all too often people procure expensive and complex AI when simpler and cheaper AI would have achieved the same level of performance, given the quality of the available data at the time, had the customer better understood how these things should be tested. It also drives the most fundamental aspect of building trust in AI amongst users.

The final chapter lays out the conclusions that can be drawn from the four cases of AI being taken from idea to actual use so far. These can form a set of real-life examples that military officers and industry can use to challenge both extremes within the hyperati, skip the trough of disillusionment through early utilisation of AI and, thus, get defence to the plateau of AI productivity before future adversaries.

Part I
Theory

2

Defining AI

Doctrine

The US Department of Defense (DoD) says that, "AI refers to the ability of machines to perform tasks that normally require human intelligence – for example, recognizing patterns, learning from experience, drawing conclusions, making predictions, or taking action – whether digitally or as the smart software behind autonomous physical systems".[1]

"Defence understands Artificial Intelligence (AI) as a family of general-purpose technologies, any of which may enable machines to perform tasks normally requiring human or biological intelligence", according to the UK Minister of Defence (MOD).[2]

That's it. Short, plain language, not at all 'techno-speak' – it's 'simples'.[3] If you remember only one thing from this book, please let it be one of these definitions, along with the realisation that what defines AI is *what it does* – specifically that it performs tasks that traditionally or normally require human intelligence – and not *how* it does it. These definitions of AI do not prescribe the nature of the inputs (such as big data), they do not limit AI to particular techniques, and they do not assume the need for extensive computing power (although, these things all help, as we shall discuss later). They do have slightly different emphases, however. The US definition helpfully illustrates the different tasks and forms AI can take. In particular

---

1 US Department of Defense, *Summary of the 2018 Department of Defense Artificial Intelligence Strategy –Harnessing AI to Advance Our Security and Prosperity* (US DoD, February 2019).

2 UK Ministry of Defence, *Defence Artificial Intelligence Strategy* (UK MOD, 15 June 2022).

3 'Simples' is the catchphrase of an animated meerkat called Alexander (with a Russian accent and a side-kick meerkat called Sergei) used in an advertising campaign for a price comparison website in the UK. It usually means something is quite straightforward. Hopefully AI will one day make such marketing silliness unnecessary.

it highlights that it can, but does not always, relate to autonomous systems such as drones. The UK definition, on the other hand, explicitly states that AI is a family of technologies and highlights the overlap between AI, machine-learning (ML) and data science.

These definitions should be considered as formal doctrine; they are published at the highest level by defence organisations after much input and consideration. Yet a large number of people within the defence community appear to be unaware of them. Worse, many working on AI in defence have said they do not agree with those definitions. A rather nice primer on AI technology published by the US Joint AI Centre (JAIC) observes that the US DoD definition "is so obvious that many are confused by its simplicity".[4]

This may well point towards a potential driver of the lack of acceptance amongst those working with AI. It is likely that what the author has labelled confusion is, in this case at least, a euphemism for the self-interest (or bias) that makes certain members of the AI community cautious about simply accepting the US DoD and UK MOD definitions. That leads to statements such as 'AI is an extremely broad field that covers not only the breakthroughs of the past few years, but also the achievements of the first electronic computers dating back to the 1940s' or 'when something is new and exciting, people have no qualms about labelling it "artificial intelligence", but once the capabilities of a particular AI are familiar, they often call that merely 'software' – even old technology can still be AI'.

The rest of this chapter is devoted to arming the reader with the knowledge needed to spot, see through and bat back against the self-interest of those who challenge these definitions (on both the pro and anti sides of the AI debate).

Simple Isn't Sexy

It is a feature of all technology that, when enthusiasts and experts talk about it, they usually focus on the most recent and cutting-edge versions and aspects. This is not a purely AI or even a defence thing. In 2004, research conducted on behalf of an investment bank into the computer memory industry discovered that the average memory chip was one-tenth the size of the most advanced chip the industry was putting into production.[5]

---

4   Greg Allen, *Understanding AI Technology* (Joint AI Centre, US DOD, April 2020).
5   Darrell Jaya-Ratnam, *DRAM Busters: The coming slump* (FOR Securities, July 2003).

This bias for the cutting edge is not surprising. It is far more interesting, makes for a better media presence and attracts more funding to talk about the latest aspect of a technology rather than the bog-standard average application. However, to base policies and development on the leading edge is akin to teaching a young person to cook by only taking them to Michelin-starred restaurants, or teaching someone to drive by only ever putting them into a Formula 1 racing car. As a result, despite AI having been around as a discipline since the 1970s (and with some of the basic AI technologies having been around since the 1940s), academics and researchers will often seek to differentiate (or denigrate) different AI techniques by talking about so-called 'true AI', and insisting that it 'requires some element of learning' (as opposed to training). In one extreme case, an individual (whom I cannot name, unfortunately) insisted that AI has to 'be capable of leaps of intuition'; this was despite 'artificial intuition' being a new and separate 'thing' at the time. There may well exist people with a level of human intelligence that barely meets these conditions, so it seems curious, and unnecessarily restrictive, that these should necessarily be required for the artificial equivalent!

Similarly, commercial organisations, when asked by potential clients for an explanation, will define AI as consisting of the few techniques they specialise in, whilst downgrading all the other techniques as 'not true AI'. Such self-serving definitions should be seen as marketing spin but, alas, many military officers are taken in by these marketing tactics. For example, following a successful pilot of a particular AI algorithm for the military, the project was 'downgraded' by replacing 'AI' in the title with the words 'machine learning'. Someone in the command chain had concluded that machine learning did not fall under AI, despite the fact that the UK MOD's defence AI strategy shows it as a subset of AI, and that the pilot project demonstrated the use of a proprietary algorithm to undertake a task normally performed by humans (specifically the members of a military board). At best, things like this are due simply to the eagerness of individuals to be seen to push at the leading edge. At worst, it can be due to individuals positioning themselves for attractive business development jobs with big technology firms when they leave the forces.

Thankfully the technology-agnostic definitions of AI given at the start of this chapter show that defence ministries are rather more resilient to this spin, and this is why it is so important for those in defence to know and accept the official definitions of AI. Anyone who defines AI in narrow terms that makes their technology seem dominant whilst denigrating others' by using phrases such as 'true AI' and 'leaps of intuition', or who emphasises the data it uses or focusses on how new / advanced it is, should be challenged with the doctrinal definitions above. In general, such a challenge will elicit two responses. The first is ignorance; but if the person proves willing to better understand the 'normal or traditional human intelligence' applied to the problem or capability for which AI is proposed, this will often lead to a better alignment of the AI technology with the requirements at hand. The second response is a challenge to the definitions, and this should be a red flag to any officer or official seeking to making a difference (as opposed to just spending funds).

Ultimately, it is the awareness of the nature of different AI techniques plus a clear understanding of the user and their needs (particularly the context of the task and how human intelligence is currently applied) that will counter commercial (or self-interested) spin and ensure rapid and effective exploitation of AI. If there is a second thing to take from this book (after the 'doctrinal' definitions of AI at the start of this chapter), please let it be this defence-AI equivalent of Occam's razor:

> *Use the simplest AI technique that delivers the tasks you require,*
> *within the constraints and limitations of the situation in which*
> *the tasks are being undertaken.*

In practice, someone has to decide up front which technique is the simplest whilst still being able to deliver the task, as there is rarely enough time and budget to start at the absolute simplest and get more complex as needed. However, this is still better (from the point of view of getting the benefits of AI) than choosing the most advanced technique that the funding allows and then finding it cannot be fully exploited without lots more time and money, which is too often the case. Or, to relate it to a phrase often heard with respect to military decision making, it is better to have the 80 percent solution in time than the perfect solution too late.

The Nature of Different AI Techniques

Within the broad definition of AI there are many different techniques. Some of the common terms thrown around about AI techniques include ML, neural nets (NN), deep reinforcement learning (DRL), expert systems, supervised versus unsupervised learning, amongst others, many of which overlap. Each has its strengths and each has its weaknesses, so it is very important not to assign the specific limitations, risks or benefits of a particular technique to AI as a whole. Of course, this is exactly what those with an interest in creating a negative or positive buzz around AI in defence often do.

There are many good summaries of AI techniques in relation to defence. The JAIC primer is one; *The Dstl Biscuit Book* published by the Defence Science and Technology Laboratory (Dstl) is another[6] (from which the Defence AI Strategy took its visual summary of AI and data science). The reason these can be characterised as good summaries is that they are mutually exclusive and collectively exhaustive (MECE), a term often used by strategy consultants such as McKinsey & Co Inc when attempting to describe a client's needs or problems in the form of an issue tree or hypothesis tree. Being MECE means that the different categories cover all the issues but do not overlap.

By way of an example, here is a poor list of types of AI techniques taken from a website that offers training in AI (not referenced for obvious reasons):

a. Supervised learning;
b. Unsupervised learning;
c. Reinforcement learning;
d. Deep learning;
e. Natural language processing (NLP); and
f. Computer vision.

This list has three overlapping issues that mean it is not MECE. The first three categories (supervised, unsupervised and reinforcement learning) relate to *what* data is provided to support the learning. The fourth (deep learning) relates to *how* the data provided is used, as deep learning techniques can be applied in supervised, unsupervised and reinforcement learning situations. Finally, NLP and computer vision relate to *where* the technique is applied rather than the learning process itself. In addition, the previous combinations can apply to each of these: deep learning techniques can be applied in a supervised learning environment to achieve computer vision.

The Dstl and JAIC each focus on a particular dimension of AI for summarising the techniques. The Dstl summary characterises AI techniques in terms of their application. The JAIC summary, a variation of which is shown in Figure 2.1, structures the techniques first in terms of the AI learning method, specifically 'handcrafted' (or manually designed) versus 'machine learning', and then breaks down the techniques further in terms

---

6 M Torres, G Hart and T Emery, *The Dstl Biscuit Book: Artificial Intelligence, Data Science and (Mostly) Machine Learning* (DSTL/PUB115968, 2019).

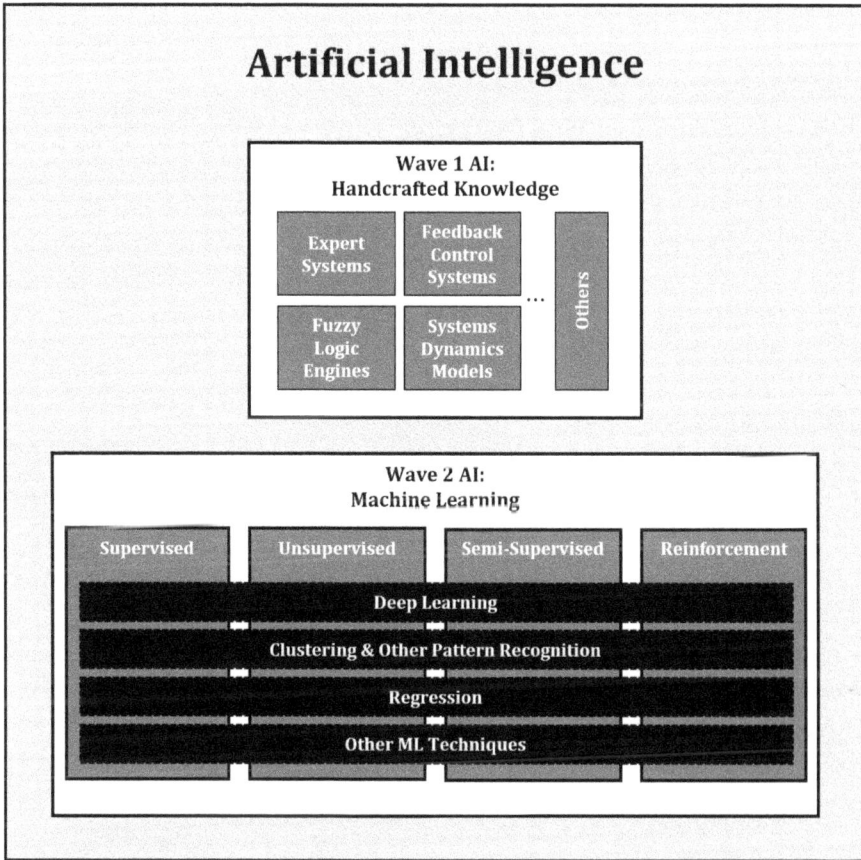

Figure 2.1: Summary of AI techniques
Source: Modified from Greg Allen, *Understanding AI Technology* (Joint AI Centre, US DOD, April 2020).

of the provision of learning data, for example experts and feedback under handcrafting versus supervised, unsupervised, semi-supervised and reinforcement under machine learning. Although taking different points of view, both highlight that ML is a subset of AI, and (by implication) that some AI does not feature any learning.

The Dstl summary may leave an AI expert a little cold by not using any current buzz-word techniques, such as large language models (LLMs), NNs or deep learning. The more recent Joint Services Publication (JSP) 936 'Dependable Artificial Intelligence (AI) in Defence'[7] has a more detailed

---

7   UK Ministry of Defence, *JSP936: Dependable Artificial Intelligence (AI) in Defence* (UK MOD, 13 November 2024).

list of the approaches and techniques that currently make up the field of AI. However, if the primary goal is for AI to perform tasks that normally require human intelligence, then understanding what task the AI is required to perform is surely the most important first step. This way of considering AI techniques has the advantage of providing a military officer charged with developing an AI capability with a simple challenge to pose to an AI expert or company that is proposing a particular AI technique: 'Explain how your (complex) AI technique is more suitable in this type of application than another (simpler) technique, given that human intelligence currently performs this task with this (limited) data'.

In the case where a specific AI technique has not yet been chosen, the JAIC representation is more useful as it supports discussions with both AI experts and potential AI users about the practical aspects. Firstly, in those (many defence) cases where data is limited or constrained, it shows that handcrafting the algorithm in the form of expert systems still counts as AI even if it is old and unsexy. This also helps to create the confidence that there is the potential to exploit AI, and for the AI user's abilities to grow in parallel with improving the data rather than waiting for the data to improve first and then catching up with the development of the AI later. However, for a handcrafted approach such as an expert system to work, one needs access to experts (obviously) and ways of extracting their expertise and converting it into an algorithm. This is not trivial, and poor expertise capture and representation has been behind many disappointments with expert system-based AI applications. However, as discussed in a later chapter, stakeholder engagement techniques previously developed for the social sciences, and often used in strategy consulting, have given expert systems a renewed relevance.

Secondly, in the cases where the user requirement is for the machine to learn, structuring the techniques in terms of the need and nature of the training data helps to understand pros and cons of the different options. Supervised learning involves gathering many data points, each with a set of inputs (for example colour, shape, size, speed) and an associated output (such as tank, aircraft); this is referred to as 'labelled data', in other words the output is labelled with the 'answer'. The AI then learns to mimic the conversion from the inputs to the labelled output with a reliability that tends to depend on the ratio of the number of input parameters to the number of data points (the more input parameters, the more data points needed to train with). The key question is whether the labelled data exists for supervised learning and, if not, whether there is the capacity (in terms of knowledge and time) for the user and/or experts to label the existing data. Many projects that have begun with much enthusiasm fail when it is realised that the data is not labelled and that the budget or timeline would not allow a labelled data set to be created. However, if a labelled data set exists then the full range of machine learning techniques become available, from regression for the prediction of numerical values, through classification using decision trees (which give categorical outputs) or naive Bayes (which gives probabilistic outputs), to techniques capable of both numerical prediction or classification such as nearest-neighbours, support-vector models (SVMs), NNs and their variants such as convolutional NN and deep NN (where 'deep' in this case simply refers to having multiple layers of NNs).

Unsupervised learning also requires large data sets, but these do not have to be labelled. The classic unsupervised learning approach is clustering and, rather than seeking to mimic the conversion of a set of inputs into a specific output for each data point, clustering algorithms seek to group different data points together based on the similarity of their patterns. The initial output from AI trained via unsupervised learning is not immediately useful; all it shows are the clusters of similar data points. However, conducting some post-processing with human intelligence, such as the humans who currently perform the tasks intended for the AI, allows the clusters to be labelled with a classification or course of action. Compared to labelling all the datapoints, labelling the clusters is very much easier and this approach is actually very similar to how humans 'pattern match' past experiences to current situations as part of the 'naturalistic decision making' approach (discussed further in the next section).

Semi-supervised learning combines an initial algorithm developed using supervised learning with a relatively small labelled-data set (the so-called 'base-learning') with insights from unsupervised learning with a larger unlabelled data set.

Reinforcement learning requires no training data set to be generated, whether labelled or unlabelled. Instead, reinforcement learning sets up 'games' with different starting conditions within an environment (specifically the inputs), chooses an output (initially randomly) and 'plays' to the end of the game. It then allocates itself a reward based on its level of success. This allocation is based on the 'reward function' that maps the reward to the state of the environment at the end of each game. The AI then learns which outputs give the highest rewards in a given situation. This is conceptually similar to supervised learning except that it generates the data sets itself randomly and, rather than labelling the data point with a desired output, generates a score for the output. A Markov decision process is a basic reinforcement learning technique, whilst DRL combines NNs with reinforcement learning. Multi-agent reinforcement learning is a subfield of reinforcement learning where a number of simulated agents learn either individually or as a group.

Not needing large amounts of labelled data is clearly a strength of reinforcement learning but it faces several other challenges. First is creating an environment that simulates the situation in which the AI is to be used accurately enough for the lessons to be relevant. Second is defining a representative reward function that is relevant across all possible starting conditions. Third is the computing power (or time) needed to learn, especially in a complex environment. Fourth, the trained AI is very complex (as it contains the lessons from thousands, perhaps hundreds of thousands, of games) and so may require significant computing power to perform tasks in real time.

A final point to consider is that an ensemble technique, using different AI techniques together, can combine the benefits and offset the disadvantages of individual techniques. In some cases, this involves running multiple techniques in parallel and then arbitrating between them. In other cases, it involves integrating the different techniques so that one feeds or guides the other.

The Context of 'Tasks Requiring Human Intelligence' in Defence

Whilst AI should not be considered just software (as the JAIC primer warns against), it is always instantiated as a software algorithm at minimum, often with an 'application' front-end. Given that concepts such as design thinking and user-centred design have been used in software development projects for decades, it is always amazing that AI developers so often make little attempt to understand the assumptions, knowledge and preferences – in other words the context – of the potential users of AI in defence.

A simple but recurring example is one of units. The Royal Navy (RN), for instance, uses a mix of units as part of their command and control (C2) and navigation processes; nautical miles for distance unless something is nearby in which case they might use cables; knots for speed except for some missiles in which case they might use Mach number; feet for altitude but metres for depth, bearing and course in degrees from north. Much of this is historical legacy but there is a good reason (or at least benefit) to maintain this, which is that the unit then indicates the context. For instance, when a RN officer in the operations room hears a figure given in metres, they know that it relates to something under the water rather than above the water, as anything to do with something above the water would have been given in feet. This is a form of double encoding, which is a technique used by human factors integration practitioners, or those involved in the design of human-machine interfaces, to minimise the chance of misunderstanding in the presentation of information. Unfortunately, whenever AI developers create an initial prototype, or proof of concept, they always use S.I. units, that is metres, metres per second, radians measured from the horizontal (the east). What is worse is that they often do not bother to put the units on the application. So, when first-time users see the application, they assume all the parameters use the units they are used to and get totally incorrect outputs. That instantly sets the cause of AI exploitation back a step as the user loses confidence due to the failure of the developers to put enough upfront effort into the users' experience and expectations. It would be easy to imagine that this is just a function of young developers having their first experience with defence, but that is not the case. As recently as the autumn of 2023, teams from two large defence technology firms, many of whose individuals had worked in defence for years, were found to have made exactly this type of unforced error. Clearly the habit formed years ago during their education of using S.I. units tend to override more recent observations of what users actually want.

More important than the units issue – indeed, the most important aspect for developers of defence AI to understand in order for the AI to be user-centric – is the context of the task or decision being undertaken. The defence context for AI is often totally different to that of commercial firms. Take some 'fun facts' about Amazon for example.[8] In 2023 it was estimated that Amazon had around 310 million active users worldwide and 600 million products listed at any given time. In a single month it attracted more than 2 billion internet visits (for instance, 2.4 billion in March 2023) and shipped 1.6 million packages a day, meaning a minimum of 48 million buying/selling transactions a month (given that some packages contain multiple orders). This is all done over a stable and high-bandwidth communications

---

8   T Bradley, "Amazon Statistics: Key Numbers and Fun Facts", https://amzscout.net/blog/amazon-statistics/, accessed 27 December 2023.

network, where it is in both the buyers' and the sellers' interest to share data on searches, products and prices. Crucially (and this is the key thing to bear in mind), all the user interactions fall into only three types: searching, buying and selling.

Now consider defence. Yes, there are more than 1 billion line items (products, put another way) within a defence enterprise. But that is split among the different services, then split further among different platforms and systems, then split further again by maintenance, training and combat usage. There is also the further split among various legacy systems with differing standards and data quality. All that complexity has to be handled in the context of networks that (at least near the front line) may not be stable or high bandwidth and may be subject to enemy action, and the sharing of the data may be constrained by security classification and national sensitivities, whilst the enemy (the other 'party') has a vested interested in not sharing data. A small proof-of-concept study using maintenance data from the primary fighter aircraft of a major member of the North Atlantic Treaty Organization found that the average number of transactions of a specific type was less than one per year. At the other end of the military process, in the nearly 50 years between the entry into service of the RN's Sea Dart missiles in 1973 (which were capable of shooting down anti-ship missiles) and the start of the Russian invasion of Ukraine in 2022, there were fewer than a dozen incidents of anti-ship missiles being fired at, or near, air-defence–capable ships.

All of this means that those developing AI for defence have to explicitly consider things that commercial AI developers can take for granted. Such defence-specific issues include the speed at which the task or decision must be completed, the computing power that can be supported, the amount and quality (however that is defined) of the available data, number of training cases per type of output, the risks of data centres or network nodes becoming points of vulnerability to an enemy strike, and the risks of AI algorithms falling into enemy hands when they capture a piece of equipment containing AI.

A counter to this view is to seek to remove many of these concerns by replicating what commerce has. For instance, the UK Defence AI Centre is addressing a range of challenges to scaling up the use of AI, whilst various UK military commands are biting the bullet and curating 'data lakes' to ensure there is sufficient good quality data for AI to exploit. In the long term (given sufficient funding over a sufficient period of time), this will undoubtedly lower the barriers to AI exploitation and benefit. However, it will take time, it will not remove these issues entirely and the enemy will always have a vote on raising the barriers again. So, it will always be a worthwhile exercise to take these contextual limitations into account, explicitly, when choosing and judging specific AI techniques.

The Parallels between AI and Human Intelligence

Given that AI is the use of a computer to perform tasks normally performed using human intelligence, it has been found useful in AI projects to understand the type of human intelligence required for the different tasks that AI is applied to.

Human intelligence has many forms. There are the primitive and reflexive actions of the amygdala, a small part of the brain that keeps you safe as a result of evolutionary processes. These include the fight or flight response, reacting to loud noises and being alert to things that look or sound like snakes.[9] Then there are the things that we are taught as children, as a form of preventive safety, which become behavioural algorithms to help us avoid having to learn through experiencing the consequences of not following them, for example looking left, right and left again, before crossing a road.

As we grow older, we experience a mix of teaching and learning in a formal setting. We are taught certain rules and methods (such as grammar, how to compose an essay, how to complete different types of calculation) and then practise applying these through exercises, tests and exams until we eventually learn to achieve the correct answers to different questions often enough to pass. Sometimes this practice and testing takes place in the abstract environment of a classroom and exam hall, whilst at other times it takes place in a simulated environment such as in military training simulators.

Finally, there are the things we learn through experience, which one might consider to be 'reflective trial and error'. Some of these experiences become reflexive, for instance not touching hot or sharp things, whilst others require memory, such as 'when I did a frontal attack I lost, but when I went left flanking, I won'. The latter is often the case when there is no single correct answer, and where there are theoretically multiple processes and paths to a successful outcome, but where achieving success depends on how well you implement any of the potential processes and paths.

---

9 Antoine Bechara, Hanna Damasio and Antonio R Damasio, "Role of the Amygdala in Decision-Making", *New York Academy of Sciences*, Vol 985, Issue 1 (The Amygdala in Brain Function: Basic and Clinical Approaches, pages 356–369) April 2003, https://pubmed.ncbi. nlm.nih.gov/12724171/.

The different AI techniques discussed previously each align with, or have parallels to, these types of intelligence, and the definition of AI encompasses them all. For instance, handcrafted expert systems are conceptually parallel to the action of the amygdala or the behavioural algorithms taught to us as children. Where the situation is obviously dangerous and the input indicators are clear, the output courses of action may also be obvious, so it makes sense to just represent them in the simplest way possible: 'if {this has happened} then do {this}'.

The many different supervised learning techniques are very much the AI's equivalent of school and college. The AI is provided with a 'question' in the form of the data sets and learns how to get the approved 'answer' represented by the labelled output. Some supervised learning techniques such as regression and decision trees are akin to learning how to answer questions in maths or physics in that, just as a student must show their work, so too can the AI show its process. Others, such as NN, are more akin to learning how to write an essay in economics or history, in that a correct or incorrect answer can be output but the writing process cannot be easily described.

Reinforcement learning is obviously the equivalent of learning by experience but with the AI able to conduct its trial and error at a much faster rate than a human. The interesting difference between AI reinforcement learning and human experiential learning is that that AI can justify its output for a particular set of inputs by displaying all the same or similar instances it has encountered and showing that the chosen output gave the best reward. With human intelligence, by contrast, it is often very difficult to get the human to explain exactly why they favour a particular output for a whole host of reasons: not actually having many similar experiences in the past, only remembering a few key features of the situation rather than every single aspect, not remembering the precise outcome or reward of the other similar situations, never having tried any other outcome, and so on. Often such things get rolled up into intuition, gut feel or subconscious memory.

The issue of unsupervised learning is looked at after reinforcement learning because it is both one of the most underrated types of AI and also the one with the most interesting parallel with human intelligence in defence. The reinforcement learning parallel to human intelligence works very well for tasks and decisions that relate to quantitative outputs such as buying and selling a house or stocks and shares, or when making business decisions. The metrics for success and failure are clear and they happen often enough that a range of solutions would have been tried to allow a comparison of different options to be made in similar circumstances. In the business world, so-called 'rational decision-making techniques', such as analytical-hierarchical process, multi-criteria decision analysis, portfolio theory, game theory and options theory, provide the framework in which human experiential learning can take place within organisations. In many defence decisions (and decisions made within the emergency services) there is neither the time nor clarity of the definition of success to consider options in this way.

Research into how decision making takes place in these situations by Gary Klein[10] led to the growth of the naturalistic decision making (NDM) movement, which provided some of the behavioural and processual underpinnings of much of Kahneman and Tversky's work on prospect theory.[11] The core to NDM is the recognition-primed decision (RPD) process. In RPDs, the human compares the current situation to their memory of similar situations. If there is a similar situation in their past experience, they choose the course of action that worked at that time. If there is no sufficiently similar past experience, humans begin to assess potential courses of action one by one and apply the first that comes close enough to the desired outcome (as opposed to going through all potential courses of action and picking the optimum, as this may take too long).

Both these stages, searching memory for the closest similar situation and going with the first course of action that is close enough, are effectively clustering processes (or pattern matching). In the first stage, the human is checking which past memory is in the same cluster as the current situation. In the second stage they are checking if the solution being considered is close enough to the cluster of acceptable outcomes. Much military decision making, particularly in the tactical space, is situation and course of action clustering against experience (or pattern matching against experience). So many observations made during trials and experiences with networks and decision aides designed to help C2 suddenly fall into place when this is appreciated: why specific experiences count (they have experiences in the same cluster), why length of experience may not (they may have many experiences in the wrong cluster, aka 'baggage'), why the best decision makers do not use the most information (only certain features of the situation

---

10 Gary A Klein, Judith M Orasanu, Roberta Calderwood and Caroline Zsambok, *Decision Making in Action: Models and Methods* (Ablex Publishing, 1993).
11 Daniel Kahneman and Amos Tversky, "Prospect Theory: An Analysis of Decision Under Risk", *Econometrica*, Volume 47, March 1979.

are critical to determine the right cluster), why experience often leads to quicker decisions (they can go with what worked last time rather than assess potential options) and so on. As discussed in Chapter 5, helping defence exploit 'simple' clustering algorithms to quickly prioritise individuals or vessels of interest, and hence better apply scarce human resources, can be rewarding.

It should be remembered, also, that many tasks that require human intelligence are part of the process of multiple decisions. That means that AI can be a small part of a process, with humans being involved before and after the AI. Thus, making the AI perform in a way that the human 'overseer' can relate to makes it easier for them to monitor, assess, trust and, if required, intervene, as they can relate to what the AI is doing in a similar way to how they might relate to a subordinate performing the task.

3

Ethics and Trust

Perspectives

> *The MOD is committed to developing and deploying AI-enabled systems responsibly, in ways that build trust and consensus, setting international standards for the ethical use of AI for Defence.*
> — UK Ministry of Defence[1]

> *To drive adoption, people need to be confident that AI is being developed and used in a responsible and trustworthy manner.*
> — KPMG[2]

> *The best way to find out if you can trust somebody is to trust them.*
> —Ernest Hemingway

It has been asserted that ethics and trust go together. Their relationship is both a practical necessity (as each drives the other) but also an observation about how often legitimate issues about ethics and trust can slow or constrain the use of artificial intelligence. This need not happen as the UK Defence AI Strategy now includes policy on artificial intelligence (AI) ethics and the Joint Services Publication (JSP) on dependable AI[3] defines 'dependable' as a synonym for 'trustworthy', so one might consider it only necessary to read,

---

1   UK Ministry of Defence, *Ambitious, safe, responsible: Our approach to the delivery of AI-enabled capability in Defence – Annex A: Ethical Principles for AI in Defence* (UK MOD, 15 June 2022).

2   N Gillespie, S Lockey, C Curtis, J Pool & A Akbari, *Trust in artificial intelligence: A global study on the shifting public perceptions of AI* (University of Queensland and KPMG Australia, 2023).

3   UK Ministry of Defence, *JSP 936 Dependable Artificial Intelligence (AI) in Defence* (UK MOD, February 2024).

understand and apply these documents. For those interested in the deeper aspects of both of these, read on.

From a software engineer's perspective, ethics and trust are important requirements to be addressed, with testing providing the evidence to help users compare, contrast and select AI. However, as with many human endeavours, from housebuilding to selling financial products, there is a tension between seeing what ethical and trust problems actually occur and then creating rules to avoid those problems from happening again versus predicting what could go wrong in theory and setting the rules early. The balance between these two extremes usually depends on the risk appetite of the relevant bodies, which in turn tends to be driven by the situation. With respect to defence technology, countries at war tend to have the former perspective of 'let's try it and see', especially if the war they are in is going badly or is seen as existential (where the existential threat could be to the country or just its government). Countries enjoying peace can afford to take the perspective of 'slow and steady', as the risk of inadvertent mistakes (and the subsequent desire to allocate blame) could be the biggest priority. Basically the underlying trade-off is driven by the consequences of a mistake caused by poor implementation versus the consequences of inaction when a military threat appears.

The case of the Royal Navy's helicopter-borne airborne early warning (AEW) radars in 1982 shows how easily defence can flip from one attitude to another if needed. The Falklands War demonstrated that the Royal Navy had a major gap in its maritime AEW capability. As a result, a design and development programme was instituted which saw the Search Water radar adapted and integrated into Sea King helicopters in 11 weeks. This is something that would normally have taken years and was possible because the authorities and defence industry were willing to take greater risks to meet the urgent need.[4] A similar and more recent case is the speed with which western defence companies carried out modifications and integration work to allow Ukraine's Soviet-era MiGs to fire US AGM-88 High-Speed Anti-Radiation Missile (HARM) missiles, and their Sukhois to fire British Storm Shadow cruise missiles.[5]

---

4   George Marsh, "Military Upgrades: How the Royal Navy Advanced Its AEW," Avionics International (1 July 2001). Accessed 25 November 2024. https://www.aviationtoday.com/2001/07/01/military-upgrades-how-the-royal-navy-advanced-its-aew/.

5   Brian Everstine, "Integration of HARM on Ukraine's MiG-29s, Su-27s Took 2 Months," Aviation Week Network (20 September 2022). Accessed 25 November 2024. https://aviationweek.com/defense/missile-defense-weapons/integration-harm-ukraines-mig-29s-su-27s-took-2-months.

This risk appetite forms the situation context of ethics and trust. This can be broken down into two levels for each: usually the level of the individual user versus the level of the system in which the AI is used. Someone raising the issue of 'trust in AI' may be referring to whether the user can be confident that the AI's outputs are sufficiently correct to act on them, similar to the way one might question whether the decisions made by a particular individual can be trusted. However, sometimes the word 'trust' is used in relation to the performance of the team or process which features AI. This recognises that the act of trusting specific AIs' outputs can impact the extent to which the overall socio-technical system can be trusted in different situations. For example, relying on the AI can lead to boredom or loss of skills which, in turn, means that failure in the AI cannot be spotted and fall-backs may not work. Similarly, questions about the ethics of AI in different areas can sometimes relate to whether or where AI should be used by an individual, and sometimes relate to how the AI is being used within the system.

As various AI projects in defence were worked through (such as those related to maritime air defence between 2015 and 2020[6]), and as issues of ethics and trust were raised and explored in parallel by different teams, the question of how ethics and trust are related to each other started to become more and more important. Are they related in the same way as the old saying about logistics and tactics ('amateurs talk tactics, professionals talk logistics')? Do they parallel the relationship between the operational and tactical levels (the realm of senior and junior commanders respectively)? Or is it just a matter of which comes first as per the chicken or egg? Philosophically one could argue that it is ethics that should come first, or that it is more 'strategic', if you will, on the basis that the ethical use of, and outcomes from, AI is more likely to drive a positive outcome for defence activity. Indeed, at the philosophical level one might agree with this. However, from a practical viewpoint one cannot make appropriate decisions on the ethical development and use of AI for specific use cases without demonstrating the level of trust one can have in the AI. All too often discussions (or, more often, arguments) are had on the basis of a convenient assumption about the trustworthiness of the AI's outputs; one side will pick a few examples of poor AI implementation, the other side will pick some good examples, and each will claim their small samples are representative.

---

6 Richard Scott, "Machine speed warfare: UK tests naval AI decision aids in ASD/FS-21 exercise", *Janes International Defence Review* (July 2021).

Let's take the famous ethics thought experiment of the 'trolley problem' and relate it to AI. The problem goes like this: There is a train or tram carriage (known in the US as a 'trolley') rolling down a track, out of control, and heading towards five people who have been tied to the tracks (for some unknown reason by some person or persons unknown). You are observing the scene some way away – too far away to release them – but by your hand is a lever that could divert the carriage down another track. However, down this other track there is one person (a railway worker perhaps) walking, too far away to see or hear your warnings or the out-of-control carriage, who would not be able to get out of the way of the carriage once it has been diverted. The choice is effectively whether you do nothing and allow five people to die or do something and cause one person to die. To a calculating and detached person, the answer is obvious: pull the lever and minimise the number of dead people. To ethics, philosophy and law students, it is far more complicated. If you pull the lever, on whose authority have you taken the responsibility for deciding who is lucky and who is unlucky? On what basis did you judge that the lives of those five particular individuals were worth more than that one particular individual? And there is a whole load of other issues that would make a great Shakespearean soliloquy.

Up to this point we have been discussing the standard trolley problem dilemma. So now we throw in the equivalent of AI trust: what if, as well as deciding whether to pull the lever or not, you also must decide whether to trust the lever? Will pulling the lever work at all? Will the act of pulling the lever at the wrong time or the wrong way cause the carriage to derail and roll sideways across both tracks, killing all six people? Might there be some electrical issue that caused the carriage to run away in the first place and might electrocute some or all of the people involved (including yourself)? Or is ignorance bliss and it is better just to assume the lever will work as intended ('trust by default')? This is where testing comes in, in order to generate the evidence on which to base your trust.

This thought experiment (which admittedly is pushed to its limits) illustrates that, from a purely pragmatic point of view, AI practitioners have to address trust first (through testing) in order to set and demonstrably achieve relevant ethical goals. As discussed later, that does not necessarily constrain ethical principles, but it does form the evidential basis for following them. However, for the military reader with potential responsibility for developing requirements, setting policy or acquisition, the place to start is with the intent of those one or two levels up the command hierarchy regarding ethics, and then explore the practicalities of trust. Testing, unlike ethics and trust, is less amenable to policy making, as it is a more technical and context specific subject. Hence the discussion of the details and practicalities of testing have been left to Part IV.

The Principles of AI Ethics

The AI Ethics Advisory panel[7] is different from the type of group that most Ministry of Defence (MOD) civil servants or defence researchers would normally deal with. In 2022, when the panel was scrutinising and advising the MOD, around five of the 15 members of the panel had first-hand experience of developing AI. One member was a serving military officer and the remainder were a mix of policy makers, academics and legal experts. Some of the discussions were highly philosophical and legalistic whilst some focussed on issues that related to the most extreme form and application of AI such as in the military 'kill chain' (even though MOD policy stated that there would always be a human in the kill chain).

The reason for this breadth of discussion was that, with the increasing concerns being raised about AI (thanks to the activities of the AI hyperati), it was necessary to explore the most extreme, worst-case issues from multiple angles in order to provide useful advice for senior officials and ministers. They also provided a good test of the scope and boundaries of the department's direction on ethics and led to the five ethical principles set out in an annex to the policy paper on the MOD's approach to the delivery of AI-enabled capability in defence ("Ambitious, Safe, Responsible").[8] Those principles support the UK defence AI strategy.[9] They are intended as a framework for the adoption of AI and are summarised below.

Human-Centricity: The first principle is that the impact of AI (and systems enabled by AI) on humans – whether MOD, military, civilians of all kinds or even the adversary – must be assessed and considered throughout the entire system lifecycle, starting with the decision to employ AI at all. Demonstrating beneficial and ethical outcomes is key to this principle and

---

7 UK Ministry of Defence, *Ambitious, safe, responsible: Our approach to the delivery of AI-enabled capability in Defence – Annex B: The Ministry of Defence AI Ethics Advisory Panel* (UK MOD, 15 June 2022).

8 UK Ministry of Defence, *Ambitious, safe, responsible: Our approach to the delivery of AI-enabled capability in Defence – Annex A: Ethical Principles for AI in Defence* (UK MOD, 15 June 2022).

9 UK Ministry of Defence, *Defence Artificial Intelligence Strategy* (UK MOD, 15 June 2022).

includes the need to consider whether not using AI would be unethical (as much as whether using the AI would be ethical).

Responsibility: The second principle is that human responsibility, accountability and control, as well as the mechanisms by which control is exercised, must be clearly established throughout the AI life cycle.

Understanding: The third principle is that the relevant humans (whether responsible, accountable or exercising some level of control) must have an appropriate level of understanding of the AI and its outputs. This must be an explicit part of the system design and appropriate to the context in which the AI is being used.

Bias and Harm Mitigation: The fourth principle is that the risk of unexpected and unintended bias and harm should be proactively mitigated. Again, this should be maintained through the development and use lifecycle of the AI, and across the full range of human stakeholders. It covers the training and testing data, the AI algorithms themselves and the actions taken on the basis of the outputs.

Reliability: The fifth principle is that AI and AI-enabled systems must be reliable (perform as expected), robust (perform across a range of circumstances or at least allow management of a situation if they end up in situations outside their training parameters) and secure. These will need to be assured regularly, given the rapidly changing nature of defence in general and the threat in particular.

A reading of the full document is encouraged, as there are lots of interesting details and nuances. However, the experience from the cases described in Part II suggest three cross-cutting points that are key. First is the repeated reference to the lifecycle. An 'ethical by design' approach helps tremendously in this area by embedding ethics thinking from the start as opposed to relying on the end user to decide on ethical issues.

Second, there are sometimes explicit and sometimes implicit references to context and appropriateness. Context and appropriateness are both the most difficult to define up front (given that military contexts are very different to the other contexts in which AI is surging) and will most impact the speed with which AI makes a difference. As the old joke goes 'the phrases "I'm sorry" and "I apologise" mean the same thing, except at a funeral'. In the same way, attempting to fix rules on how these principles are implemented for different AI will either constrain the AI to the point of ineffectiveness or will have to be so loose as to be pointless. Rather, these principles must form part of the process of design, development and assessment of the balance between risk and benefit in the relevant context – ideally from the start of the lifecycle through to use.

Finally, there are the explicit and implicit references to things being demonstrable, in other words providing evidence of the AI's outputs and performance. Too often issues of generating evidence through testing are left to the AI developers, and the military sponsor or user has to try to catch up with understanding these results. Inevitably there is insufficient time and resources for them to fully understand, and even less time and resources to redo any testing that is found to be unsuitable. Yet evidence and understanding are key to building stakeholder trust in the AI.

Dependability

Trust is an emergent property of the confidence a human has in someone or something and consists of four elements which are common to humans and AI applications. First is the trustworthiness, or dependability, of the decision-making agent be it a human or AI application. This is the thing that AI developers can directly improve. Next comes the experience that one has with the human or AI application which allows a user or organisation to observe and accept the level of dependability it has. In other words, if you do not see it being dependable or trustworthy you cannot trust that it is. AI developers cannot affect this directly but those responsible for running the procurement and implementation programmes for AI can ensure this takes place. Another element is the ability to interrogate and confirm the confidence that the human or AI decision-making entity has in its own outputs. In a normal team and command practice, a commander can ask a subordinate "are you sure?" or "what's your confidence in that?". The human decision maker can then say "I'm sure" or "actually I'm not sure" and the commander and their subordinate can then assess whether the confidence is proportionate to the situation and whether to go through the decision-making process again or go with whatever the current output is With a human teammate, being honest about their confidence is a key trust builder. Saying "I'm not sure" is perfectly acceptable as long as it is not a constant issue and certainly far less of a problem than being overconfident. The final element of trust is that the human or AI decision-making entity is surrounded by some system and process that can check their working and catch any errors.

Dependability is the foundation of trust, and a major driver of ethical principles partly because it is the one element that can be directly affected and tested. Following the publication of the AI strategy, and the 'Ambitious, Safe and Responsible' policy paper, the MOD published 'JSP 936: Dependable Artificial Intelligence (AI) in Defence'[10] as a first step in implementing the principles, with the word 'dependable' adopted as an overarching term for trust, trustworthiness, reliance, robustness and confidence. As with the other strategy and policy papers referred to in this book, this JSP is thoroughly recommended reading but the key points are as follows.

Definition of AI: It provides two definitions of AI which move forward from those given in the UK AI strategy. It states that AI can be characterised as "machines that perform tasks normally requiring human intelligence, especially when the machines learn from data how to do those tasks" and "technology designed to approximate cognitive abilities including reasoning, perception, communication, learning, planning, problem solving, abstract thinking or decision making."

Types of AI systems: It classifies the two types of systems in which AI operates. The first is robotic and autonomous systems (RAS), which are typically physical platforms that deliver an effect outcome. The control loop for RAS can, potentially, be closed, meaning it is fully autonomous or could have a human in or on the control loop to oversee the AI and take over if needed. The second type is digital systems, which are provided with data and offer an output for use by a human or by another system. This class might be more aptly called 'digital decision aides' or 'digital decision systems' as normally that is what they are used to inform or to support. These tend to be open loop, in that they must be linked to another system or a human before any effect or outcome is achieved.

Proportionality: The JSP makes the excellent point that dependability relates to how much the user can rely on the AI within a particular context, and that this means the level of confidence must be proportionate with the reliance placed on it. It goes on to highlight that the evidence to support this confidence must be judged by whomever owns the risk at each part of the lifecycle, who must also communicate this risk to stakeholders (particularly users).

The JSP does not detail how to measure and prove dependability or provide a scale that links dependability to reliance on the AI. It does however provide the high-level framework in which the necessary activities can be undertaken to, ultimately, drive trust in the AI.

---

10 UK Ministry of Defence, *JSP936: Dependable Artificial Intelligence (AI) in Defence* (UK MOD, 13 November 2024).

Great Trust Expectations

Why do we trust anyone or anything (not just AI), and what evidence of dependability do we actually have and use to base this trust on? At the most basic level there are three reasons we trust someone or something, outside of a personal relationship. First, we just have no choice. We know next to nothing about the driver of the train we are on or the pilot of the passenger plane we are in but if we want to get to where we are going we have to trust them – or, if not them as individuals, at least the system which hires, trains and managers them (which we also are likely to know nothing about and have no choice but to trust). Second, we have no particular reason not to trust them as we have not seen or heard anything that indicates something could go wrong, how frequently this happens or how bad it could be. If it has never occurred to us to question the trustworthiness of the train driver or pilot, then the default is passive trust. Third, we have evidence to hand that something is trustworthy and dependable, assuming of course that those who generated the evidence are themselves trustworthy and dependable. One presumes they would have evidence of that too!

When people in large organisations talk about trust in AI they are referring to the assumption (implicit in the MOD and KPMG quotes at the start of this chapter) that we must have evidence that something should be trusted before we can get to the point where we accept we have no choice but to trust. It may be observed, however, that the rapid rise in the use of ChatGPT and various deepfake technologies suggests that Hemingway's view that the best way to find if someone can be trusted is to trust them is more common in practice than people care to admit or are willing to accept as a formal or explicit strategy. On the other hand, there is an interesting contradiction highlighted in the KPMG study, which states that 76 to 82 percent of respondents have confidence in defence organisations "to develop, use and govern AI in the best interest of the public" whereas one-third of respondents "lack confidence in government and commercial organisations to develop, use and regulate AI".[11] The contradiction is that (trusted) defence is made up of (less trusted) government and commercial entities, and this makes the drive to seek evidence a potentially useful thing to do.

The problem is that new technologies often suffer from an expectation of higher risks or higher standards than whatever they are seen as replacing

---

11 KPMG, "Trust in Artificial Intelligence: A Global Study (2022)". Accessed 25 November 2024. https://ai.uq.edu.au/files/6161/Trust%20in%20AI%20Global%20Report_WEB.pdf.

(at least until they have the necessary observed experiences), and this can slow their adoption. Take, for example, the motor car or 'locomotives on the highway' as they were initially referred to by the authorities. Between 1865 and 1896, the maximum speed limit was two miles per hour in a populated area and four miles elsewhere, limits that were lower than the average walking speed of a horse, which is four miles per hour. In 1896, the speed limit was increased to 14 miles per hour and then to 20 miles per hour in 1903, which is significantly lower than the 37 miles per hour that is a horse's maximum speed.[12] The current proliferation of 20 miles per hour limits in cities such as London is the result of more recent concerns about car pollution and risk to pedestrians through high population densities, which were not issues in the early 1900s.

Biases in Trust versus Dependability

Given that AI is a decision-making technology, what level of decision-making ability counts as 'trustworthy'? In a survey conducted across 31 countries in 2023, doctors were considered the most trustworthy professionals. The highest levels of trust were in Spain, the Netherlands, Indonesia and Argentina, where 68 percent of the people surveyed said they trusted doctors (in the UK and US the figures were 62 and 58 percent respectively).[13] What is of interest is how this level of trust relates to the decision-making ability of doctors.

Given the significant amount of time medical students spend at university, exam performance would seem a good indication of decision-making ability, at least in the sense that getting an exam question right is a proxy for getting a diagnosis right. A number of different sources indicate that 92 to 98 percent of medical students in the UK get a first or upper-second honours degree, in roughly equal numbers.[14],[15] At first sight this sounds reassuring until it becomes evident that the students are awarded an upper-second if they score between 60 and 70 percent on their exams. That means that 50 percent of the doctors make decisions about patients' health despite

---

12 Counsel Direct, "A Brief History of Speed Limits" (8 February 2015). Accessed 31 March 2024. http://www.counsel.direct/news/2015/2/8/a-brief-history-of-speed-limits.
13 Statista, "Trust in Doctors Worldwide by Country". Accessed 31 March 2024. https://www.statista.com/statistics/1274258/trust-in-doctors-worldwide-by-country/.
14 University College London, *Honours Degree Outcomes Statement 2019/20* (UCL, 2020).
15 Higher Education Statistics Agency, "Table 50: Students by Subject Area and Sex". Accessed 31 March 2024. https://www.hesa.ac.uk/data-and-analysis/students/table-50.

getting 30 to 40 percent of their exam questions wrong. As for the half who get a first, most get between 70 and 80 percent, which means a majority of the 'better half' of medical students may still get 20 to 30 percent of their answers wrong. Would an AI application that was shown during testing to be wrong 20 percent of the time be considered trustworthy enough to be relied on? Yet we appear willing to accept worse from medical professionals.

Now consider how managers and executives – people who use their human intelligence to make decisions (as opposed to making things) – are told to progress. The most common advice recruiters give candidates is to be confident, not to make sure they're good at what they're applying for. A quote by the British entrepreneur Richard Branson that did the rounds some years ago was "If somebody offers you an amazing opportunity but you are not sure you can do it, say yes – then learn how to do it later!" And then there is the famous 'fake it 'til you make it' motto of much of the tech sector. These are examples of the normalisation and exploitation of a cognitive bias known as the halo effect, in which people assume that confident (and good looking) people must be intelligent and competent.

That the assumed link between confidence and competence is a bias (at least in decision making) is illustrated by the results of an experiment on confidence versus correctness amongst doctors and weather forecasters, published by Professor Scott Plous.[16] Both groups of professionals were given relevant information to allow them to make a decision; the weather forecasters received data on recent weather patterns and were asked to predict the weather, the doctors were given case notes and were asked to make a diagnosis. Crucially, they were also asked to indicate their confidence in their decision. Depending on one's point of view the results were either intriguing or terrifying.

In the case of the weather forecasters, their confidence and correctness were almost perfectly correlated. When they were 90 percent confident in their prediction, they were around 90 percent correct, and when they were 50 percent confident they were around 50 percent correct. By contrast, the doctors had very little correlation between their confidence and correctness; when they were 90 percent confident in their diagnosis, they were actually correct around 15 percent of the time, and when they were 50 percent confident they were actually only correct about 10 percent of the time. Behavioural

---

16 Scott Plous, *The psychology of judgement and decision-making* (McGraw-Hill, 1993).

analytics research undertaken in other sectors since then[17] has suggested that the key driver of the reliability of weather forecasters compared to doctors, in terms of accuracy versus confidence, is due to the much shorter and clearer feedback cycle of whether they were right or wrong. Weather forecasters get to see whether their predictions are right or wrong quite quickly whereas doctors, once they have diagnosed and referred their patients, may not know for some time if at all.

In many (perhaps most) AI projects, users ask for some indication from the AI of how confident it is of its output. This would seem useful but, given how humans misuse indications of confidence to get ahead, it should be considered whether those who produce AI might end up falling into the habit of creating halo effects for their AI. It is often the case in defence that the 'new and shiny' technology (often referred to as 'Gucci kit') can get more attention and funding than potentially more useful things. Some years ago, defence organisations preferred to fund neural-net solutions to problems where simple rule sets would have worked just as well or, more recently, preferred to spend large amounts of money on LLM based applications when an inexpensive chatbot would have solved the problem more quickly. So perhaps AI developers might be tempted to use attractive avatars or exaggerate the confidence level of the outputs to win contracts rather than focus on the actual correctness of the AI decisions.

System Checks, Challenges and Catches

What has not been presented regarding the correctness (dependability) or lack thereof of decision-making entities, such as doctors, is that they are part of a system that will usually have processes in place to check and challenge individual decisions and catch any errors if needed. Similarly, any AI used within an organisational process would face checks and controls. Hence, trust should not be about evidence of performance of the AI in isolation (which is the dependability element of trust) but about the performance of the system once the AI has been integrated, although clearly the dependability of the AI is the starting point. After all, the military can provide 16- to 21-year-old youths with somewhere between a few months and a couple of years of training, give them a lethal weapon system, and trust them not to kill civilians or their comrades. That trust is given before they have provided any

---

17 James Montier, *Seven Sins of Fund Management: A behavioural critique* (Dresdner Kleinwort Wasserstein, November 2005).

evidence of how they will cope in combat because there is trust in the system within which these 'beta versions' of human intelligence (to use the software metaphor) have been deployed.

For these reasons, the conclusion is that trust in AI should be built in the same way that trust is built amongst people: through qualification, regular testing and confidence reporting within an 'experience management' system (an equivalent of the 'career management' system of the armed forces). This view is not widely shared within defence. The prevailing attitude seems to be that AI should be subject to one of two existing approaches. One is to assure AI in the same way that software is assured (particularly safety critical software) on the basis that AI is instantiated in some form of software. A good outline of this approach is the *Dstl Biscuit Book: Assurance of AI and Autonomous Systems*[18] which incorporates trust and ethics. The other is to validate the AI in the same way as models and simulations are validated before they can be used to make business-case decisions by large organisations.

Both the software assurance and model validation approaches generate some evidence that could drive trust, but both have significant limitations for certain types of AI and AI applications. Software assurance is needed to ensure that the implementation of the AI achieves the maximum benefit from the AI's underlying performance. Validation is needed to confirm what that underlying performance is, and to confirm that the overall application achieves the same level after implementation.

The limitation both approaches have in common is their dependence on the consistency and predictability of outputs. To assure and validate there needs to be a pre-determined correct 'answer' which the AI needs to output when given a particular set of inputs. For the earliest AI technologies, such as expert systems and fuzzy rule sets, this works well enough as the systems are developed to mimic the views of the experts. The main challenge in these cases is to demonstrate that the assurance or validation activities have covered a sufficient proportion of possible input combinations and can reliably give the pre-determined correct answer each time.

However, many AIs (like most humans) learn and apply seemingly random variations to their outputs, which would create the impression of being unpredictable when faced with assurance approaches (as opposed to

---

18 UK MOD "Assurance of Artificial Intelligence and Autonomous Systems: a Dstl biscuit book", last modified 1 December 2021, https://www.gov.uk/government/publications/assurance-of-ai-and-autonomous-systems-a-dstl-biscuit-book/assurance-of-artificial-intelligence-and-autonomous-systems-a-dstl-biscuit-book.

being given the label of 'creative' that we might apply to equally unpredictable humans). If these AIs are applied to problems which human intelligence struggles with, then validation would also not be appropriate because the whole point of the AI is to come up with a better result than humans, not mimic them.

Being able to deal with learning, and problems which do not have obvious answers, is where the experience and qualification applied to humans are more useful. The process is as follows: provide intelligent entities (of whatever kind) with some training, qualify that they have learned the core curriculum of the training through testing, allow them to apply this learning to various experiences, review how this application of their learning went in terms of performance, and then use the feedback to update their internal 'algorithm'. This translates equally well to AI as to humans if it can be accepted that AI and human intelligence are just decision-making systems with some level of unpredictability and creativity.

Part II
Practice

4

DUCHESS: AI That Captures the Lessons Learned by Human Intelligence

Learn from yesterday, live for today, hope for tomorrow.
The important thing is not to stop questioning. – Albert Einstein

Key Takeaways

The DUCHESS story is an excellent case study of what happens when the technical fails to take account of the organisational issues. It was initially funded and used by defence organisations but to fulfil a function that many commercial organisations already made use of (albeit without AI). However, as the application developed over the years with advances in the underlying technology, defence organisations found it harder to use than commercial organisations did. This is not a criticism. It is an observation of how the defence context features constraints that have to be considered by those developing AI for defence.

The first four steps of the utilisation staircase described in Chapter 1 proved easy to address. *What existing (human) process is it addressing or similar to?* DUCHESS was used to conduct face-to-face lessons-learned interviews. *What is the difference to other AI?* Unlike traditional chatbots with rigid question trees, DUCHESS listens and generates tailored follow-up questions in real time. *What does the AI do better, or allow one to do, that was not possible before?* It reveals the unknown unknowns by conducting interviews at scale where hundreds of personnel could talk openly without fear of judgement, repercussions or rank deference. *Is it as complex as needed (or simple as possible) or as complex as possible?* Simplicity won out; the interface was designed for click-and-talk ease, and the AI avoided unnecessary technical flourish. Things went well whilst the development team were focussed on the underlying problem and functionality.

The key takeaway is that the biggest driver of adoption was less about the technology or performance and more about the ability of organisations (or perhaps the freedom given to people in those organisations) to take ownership of the application and overcome the inevitable technical issues. Although there were technical answers to address the questions regarding realisation and trust building (perceived versus actual risks, time to mature and integrate, ease of use), the answers were difficult to prove. To complicate matters, the answer to the question of whether the developers really understood how the military user would use it yielded far too many positive answers. Alas, the greater the potential use across an organisation, the less able one department was to take on the responsibility for the AI. Ironically, the national and international defence organisations who have made the most use of DUCHESS are not those who provided the initial impetus.

Context

Innovation and technology can throw up some interesting ironies. A favourite is that computers now ask humans to prove they are not robots on websites. Another is the jest that before we worry about artificial intelligence, we should address natural stupidity (a quote by Guillermo Del Toro[1]). The irony here is two-fold. First, it takes individuals with a fair amount of natural intelligence to create the artificial variety and, second, there is an awful lot of natural intelligence that is simply left untapped. The current and rising interest (or concern, depending on one's viewpoint) in large language models (LLMs), and generative AI is a case in point. Organisations are spending a lot of time and effort working out how to get AI to answer their questions through, for instance, 'prompt engineering'. Yet too often they overlook those at the coal-face of the organisation who, in many cases, have the answers already. Even more importantly, it is usually those at the coal-face who have the best idea of what questions to ask in the first place (noting Einstein's various quotes

---

1   Speaking at the 2023 Toronto International Film Festival Visionaries event, Mexican filmmaker Guillermo Del Toro said, "People say 'are you worried about AI?'. I'm worried about natural stupidity".

on the importance of determining the correct questions rather than running to provide an answer).

However, even those organisations that realise their people are a rich source of insight struggle to get the answers they need, particularly with regard to the fundamental problems they face. There are two reasons for this. The first is that automated surveys and chatbots allow access to lots of people, even if they are geographically dispersed, but can only ask questions about things or problems that the operator has already identified. That is because surveys and chatbots rely on preset question trees. Yes, they might include a free-form option to fill in 'anything else you'd like to tell us' at the end but by this time most people are in the 'direct answer' mode of thinking and have been led (perhaps even biased) by the nature and tone of the previous questions.

Typically, when an organisation needs to get some rich or deep information it has to invest in some kind of face-to-face structured interviews or workshops. Yet in many organisations these exercises do not get to the real problems either, for a host of interpersonal reasons. The person leading the interview may have biases or hobbyhorses, so there is a good chance they may focus on issues they are already interested in. That person may also have been biased by the early interviews; if the first person mentions something of interest, the interviewer might be tempted to ask all the subsequent interviewees about that issue, even if the interviewee is not actually interested or knowledgeable about it. Then, when the interviews are analysed, that particular issue becomes a common theme for no better reason than it was brought up early and referred to by the interviewer in a way that led the subsequent responses.

Many problems can only be uncovered if the interviewees are willing to raise issues about, for instance, their own and others' competence and capability or when they were forced to ignore procedure in order to get things done. In many organisations, deference to rank or superiority may motivate people to give the answers that they think are expected, to make their seniors look good, or to leave out those problems that appear to have been dealt with (which often means someone else has to deal with them again). Finally, they might fear being seen to waste time and so give reflex answers rather than considering the question deeply and going into the detail.

DUCHESS was developed to replicate the nature of a human face-to-face 'lessons learned' or 'issue exploration' interview, which involves listening to the interviewee and generating probing questions based on what is said rather than having a fixed set of questions. But, being AI and delivered online, it has the benefit of surveys in terms of reach and scale, plus the benefits of drawing out the truth by not biasing the questions and not judging or intimidating the interviewee.

Inspiration

The push to develop what turned into DUCHESS began because a colleague, Sarah Vincent-Major, and I each had experiences of both the benefits and difficulties of engaging with large numbers of stakeholders to understand organisational problems. My experiences concerned strategy and technology consulting whilst Sarah's related to how the military attempts to learn lessons.

My firm, DIEM Consulting, was rooted in the operational analysis techniques I learned when working at the Defence Evaluation and Research Agency combined with the consulting techniques I learned from strategy firm McKinsey and Co Inc. A core part of the 'McKinsey way' was the information-gathering interview, normally conducted with quite senior people in an organisation (as McKinsey prides itself on taking on board-level issues). We would generate a list of specific questions we knew we needed to answer, group them into themes and then come up with a good 'open question' that covered each theme. The idea was that we would ask an open-ended question and then let the interviewee just talk, on the basis that people (especially senior people) love to provide others with the benefit of their experience. They would naturally answer many of our specific questions, but they would also address issues that we did not know we needed to know, and hence would never have thought of asking. Usually, once we had got through all the open questions, we had very few of the specific questions left (which we could ask at the end) and a lot of extra relevant information.

The DIEM method varied in one respect. During the preparation we would create some kind of visual representation of our understanding of the problem (usually in a causal map or influence diagram) for the interviewee to react to. We found that providing less senior people, who tended to know more about the issues at the coal-face but might be inhibited by what the higher-ups might expect them to say, with a picture to criticise and correct made it easier for them to raise difficult issues. After all, they were not criticising anyone in their company, only the picture presented by some external consultants (us). It also meant that as we updated the picture based on their views and experiences, they were far more willing to buy into, and act on, our recommendations because they could see how we had incorporated their inputs. This, in turn, meant that despite being a tiny company, we developed a reputation for producing outputs that helped get things done.

The expensive part of all this is the time it takes to interview a suitable number of people. At that time it was all face-to-face interviews, so we had to get time in their diary, travel there, do the interviews, take notes as close to a transcription as we could without actually recording it, write up those notes and then use them to develop the causal map. All this was very time consuming and hence very expensive. For McKinsey the cost of the interviews was not an issue, whereas for us the interviews were costly but they provided evidence to support future requirements that meant it was considered a 'spend to save' measure.

Sarah's experience came from when she worked at the Maritime Warfare Centre (MWC). One of the MWC's many responsibilities was to capture lessons learned from Royal Navy (RN) ships once they had come back from a deployment, but only the commanding officer would have a face-to-face interview. The rest of the personnel on board may, occasionally, have been asked to complete a survey but, generally, did not get asked anything. This meant that the formal written records of that deployment were based on the experience of the commanding officer, perhaps taking into account information fed to them by the heads of department. On a frigate or destroyer, there are 200 other people that did that same deployment who would have had different experiences, but those were never captured.

The key thing (and why it is important to capture everyone's experience) is that everyone would have come across problems or issues that they had to find ways to solve. But they would only have gone up their chain of command if it was important enough to bother the boss with, which normally only occurs if they could not solve the problem. If they managed to deal with it themselves, then problem solved, they don't need to bother other people with that, the boss doesn't need to know – they've got enough things to worry about. Alas, this attitude does not help the lessons-learned process, as all those solutions and good practices that develop are not captured. So when the next ship goes out to do the same deployment and hits upon the same problems and the same issues, its crew once again has to find ways to solve them. They cannot take the experience of the people who went before them. So it repeats the process, which is inefficient, can be demoralising and detracts from doing more useful things. So it can be a real issue if only the person at the top is interviewed, yet this, of course, is what happens because time and resources are limited. The time it takes for an interviewer to conduct face-to-face interviews with 200 people means it is just not going to happen at the end of a deployment.

Even if everyone is interviewed, something has to be done with all the information gathered. Does it get recorded and transcribed, and then is it analysed? To find trends across ships or between different ships, some analysis will be required, but this is also time consuming and may be prohibitive in a resource-constrained environment. Yet without a consistent approach and regular cross-checks and comparisons of the lessons from across the ships, the exercise has no value. Whereas the costs of conducting and analysing face-to-face interviews can be borne as part of a strategy or research project once every few years, the costs of doing it for every ship and every deployment soon mount up.

These two sources of inspiration came together in 2017. We happened to be presenting our strategy work to the head of the RN's innovation cell, a serving captain. Our presentation focussed on showing how some of the causal maps we had developed had been used by large organisations, including parts of the Ministry of Defence (MOD), to prioritise investments in technology research and development. Causal maps show the linkages and causality within a complex problem space, and can be an effective visual representation of how the many parts of a problem space fit together. The captain asked us to develop a causal map of her problem space, which concerned innovations to solve problems within the RN. Generally, no one had complete knowledge of the overall problem space so the fact that we developed these causal maps via interviews with people who had expertise and knowledge of each part of the problem space was a key attraction. Interviewing a number of subject matter experts (SMEs) meant we were able to build the picture, see where their knowledge overlapped, and use the picture as a basis for the discussions; ultimately the models needed to determine cost-benefit. None of this was amenable to simple surveys but the process of conducting the interviews was expensive. In 2017, it cost around £50,000 to conduct studies using our SME interview approach due to the time and effort needed to organise, conduct and analyse the stakeholder interviews. This led to the head of the innovation cell telling us she loved what we did but even £5,000 was beyond reach. It was clear that doing something for the RN innovation cell was not feasible at that time. However, Sarah and I were keenly motivated to find a way to reduce the cost.

We knew that we still wanted to interview people to get rich narrative insights to help create the causal maps of the problem space, but could we automate the process to reduce the time and resources needed? Could we come up with a way to use a chatbot to conduct the interview in a conversational way rather than simply asking a set of survey questions? Could we take that narrative data and automatically analyse it to get some insight with minimal human effort?

Inception

DUCHESS was just an idea, perhaps nothing more than a desire at this point. I was lecturing part time at Birkbeck College, University of London, and one of my students introduced me to the use of natural language processing (NLP) techniques to analyse narrative documents, specifically regulations in the financial sector. On investigating this further we concluded that NLP was something we might use but it was at a very low 'technology readiness level' at that point. This was years before ChatGPT, so we needed to try and get it funded by someone with deeper pockets than our own.

At the time there was the Defence and Security Accelerator (DASA), which was the UK MOD's innovation fund into which any company could put ideas and, if the MOD liked the idea enough, they might fund it whilst allowing the company to keep ownership of the intellegctual property. So DIEM wrote and submitted a proposal with the initial use case being for lessons learned interviews for ships coming back from deployment, inspired by Sarah's experience at MWC.

Having submitted our proposal into the 'open call' (which welcomed ideas on any subject), and before we had been notified whether we had been successful or not, the MOD launched a competition looking for innovative ideas specifically focussed on 'people in defence'. One of the focus areas of this competition was morale and retention, which we thought was a major aspect of our proposal, that is being able to interview people could contribute to making them feel like they have been heard, that their knowledge is considered important and is worth sharing, and that someone wants them to share that knowledge. In other words, it might make them feel more valued, which might have an impact on morale and retention. So we resubmitted our proposal to that competition. In the first of several successful proposals, we got funding for phase 1, which was to build a proof-of-concept application.

Having got some proof-of-concept funding, one of the first problems that we had to address was that, in defence, much of the data generated is sensitive and ought not to be flying around the internet. So we had to find a way to do the speech-to-text transcription element without using online connectivity. This was a real challenge. Most good speech-to-text engines rely on connectivity, because the other 95 percent of applications, maybe even higher than 95 percent, have no worry about being online. So why would companies and individuals develop things that were going to be used by only a very small part of the marketplace? We tried a number of different speech-to-text engines, which had varying levels of performance. In the end we settled on one of the leading dictation engines, which was designed for people sitting at a desk dictating into documents or custom forms.

The second challenge was dealing with the amount of data from the interviews needed for any analysis to be useful. The technical idea underpinning DUCHESS was that we would ask an open question, the interviewee would answer, the response would be analysed and that analysis would lead to the generation of a probing follow-on question relevant to the individual interviewee and what they had just said. However, a one or two sentence response does not generate enough data for the NLP techniques we were using to come up with good quality follow-on questions.

We addressed this problem in several ways. Initially we developed some follow-on questions that did not pull any content from the interviewee's initial response. These were not probing follow-on questions but, rather, generalised or generic questions to get them to speak for longer so that more data would be generated on which the NLP analysis could work. We basically determined the minimum word count needed to generate probing questions of sufficient quality, and if the response contained fewer words, then there would be a generic question. We had not planned to do this; it was simply the result of testing and seeing how much data was actually needed for these early AI techniques to be effective.

Another way we addressed this problem was to improve the wording of the initial open questions. We discovered that there was an art to writing these questions in such a way that they are not leading and are genuinely open, and also encourage people to talk for a long period of time. Of course, if they have nothing to say, they could just skip a question, but getting those initial questions right proved a crucial, but non-technological, challenge that we had to overcome to encourage people to speak.

Another challenge was doing the analysis once we had the interview transcripts. We wanted to machine-create a causal map automatically from the interviews, as this was something that we had been doing as humans, to help meet our aim of doing it for £5,000. This proved really difficult. One issue for the machine was being able to work with imperfect data. It did not have the contextual understanding to correct or interpret transcription errors, as LLMs are now able to do. We often got gobbledygook and struggled to make progress with the automated causal mapping.

So we decided to focus on the analysis that we could do with the NLP technology we had at the time. We could do frequency analysis to look at what people were talking about a lot. We could do sentiment analysis. We could highlight what were people talking about that was important to them. Over time, we have added to those initial analyses, and are now able to look at the spread of sentiment towards different topics versus the sentiment of individual interviewees. For instance, are some people just generally really positive, are some people generally negative or, actually, across the questions, do people have quite mixed views in terms of their sentiment towards things? In those early days, it was risky to de-scope the analysis due to the challenge it posed compared to what the technology was capable of delivering, as we did not know if what was possible was useful to the customer. Yes, we could deliver lots of stuff (interview transcripts, supported by frequency, sentiment and importance analysis), but would that be helpful? Our starting assumption was 'probably not' for our intended 'lessons learned' audience.

Technology

DUCHESS evolved in terms of the techniques that we used. When we first started we were using dictation engines that relied on the person speaking to say the punctuation. Clearly, we could not get people doing post-deployment interviews to finish their sentences and then say 'full stop', so we had to come up with a way of getting the machine to punctuate the spoken text automatically. This was important partly because the text needs to be readable but mainly because an analysis of sentiment towards topics or subjects generally works on a sentence-by-sentence basis. So sentences need to be defined accurately in terms of where they start and end. Therefore we applied text segmentation techniques to the output of the dictation engine.

As time went on, we built further bespoke algorithms that combined methods for punctuating the text with techniques such as 'parts of speech' tagging to support the sentiment analysis. This tagging identified what type each word in a sentence is, to be able to identify the sentence subject. For example, if people are talking positively about something, what is the something within that sentence that they are talking about? This led to us developing a sentiment analysis algorithm using, as a base, a sentiment lexicon which is an extensive list of words coupled with a sentiment score. So, the word 'fantastic' would have a high sentiment score, whereas the word 'appalling' would have a low sentiment score. This was then coupled with rules concerning the language around those words, so 'it was less than fantastic' or 'it was not great' would get a very low sentiment score. The word 'not' is very important when used in conjunction with 'great', because otherwise if you just take the word 'great' in isolation it would seem like people had a really positive level of sentiment towards the subject of the sentence.

We ended up with a combination of semantic, sentiment and frequency analysis algorithms within the NLP engine that worked out what probing follow-on question should be asked next. The core logic was a comparison of the extent to which the person had spoken very positively or negatively about a subject compared to how much they spoke about something. In a human face-to-face interview, if someone makes a very positive or negative statement in passing about something, the interviewer would ask a probing question to draw out more detail. That is what we were trying to get the machine to replicate when asking a follow-on question: 'I want to know more about [that thing that you haven't spoken about much, that seemed to be important to you]'.

The logic tree we developed to determine the nature (as opposed to content) of the next question has prevailed throughout the life of DUCHESS. This indicates whether the next question should be a generic one or a probing one, which sentence it should use as the basis for the follow-on question, and what content from that sentence it should put into the probing follow-on. We have refined the techniques that we use, from bespoke combinations of NLP techniques to the current advanced LLMs, but it is still that same logic of trying to get them to expand on something that seemed to be important to them but have not talked about very much, just as a human interviewer would.

The test of DUCHESS was the extent to which it had the flexibility that a human interviewer had of being able to listen to the interviewee's answer and to ask questions based on what the interviewee has just said. We had to be able to demonstrate the contrast between this and surveys which have a set list of questions, or chatbots that have a look-up table of questions to ask based on preset 'trigger' words or phrases ('if they mention X then ask them Y'). It is worth saying that some commercial chatbots appear very conversational, which makes them seem as if they are listening and being flexible with their questions. However, this is just a function of the maturity of chatbot design and the fact that they are used for specific interactions, such as between customers and call-centres, in which large amounts of past data can be used to train models to know what to expect and have a response. The whole point of DUCHESS interviews was the ability to ask an open question and allow the interviewee to talk about whatever was important to them, and then ask probing follow-on questions with no prior knowledge of what they might talk about.

Proof of Concept

When we came up with the idea of DUCHESS, the main purpose of using AI was to avoid having a human interviewer, which would reduce the cost of conducting the interviews and also reduce the risk of an interviewer biasing the questions.

We also had a supporting hypothesis that an AI interviewer would give people the freedom to say what they wanted to say, or felt they needed to say, without facing human judgement. This was based on our previous experience that carrying out interviews with the military can lead to all sorts of behaviours that are typical of working in a hierarchical organisation. People give the answer they think is expected or wanted of them rather than sharing their true feelings, because they do not want to damage their relationships with the people that they work with. Someone on a ship for six months does not, necessarily, want their shipmates to know their innermost thoughts about the management on board or how people are doing their jobs. Moreover, following on from the risk of damaged relationships, there can be a perception that being too honest can actually hamper career prospects.

Balancing our view that an AI interviewer would allow interviewees to be honest without risking their careers, we had a third hypothesis that an AI interviewer would not be able to replicate a human entirely. In particular, we suspected that the quality of the questions would probably not be as high as those which a human would come up with, but we thought (or hoped) they would be good enough to generate more insights into the 'unknown unknowns' than surveys would.

The proof of concept was tested with the crews of two small ships that had recently returned from deployments, followed by individuals involved in a headquarters transformation project, and the results related directly to these hypotheses. The first thing we discovered was that our hypothesis that people could and would be more honest was true. A common response, across the various ranks and levels of seniority, was that respondents did not have to give the answer that they knew (or at least thought) management would have wanted to hear. They could give their true views without worrying about wasting an interviewer's time. It is quite a profound comment that someone junior may think their views are less important because they are junior yet a ship could not function without each person playing their unique part. Hence their views and experiences may turn out to be just as important as the commanding officer's. People also commented that if they were interviewed by a human they would worry about wasting their time if they did not have anything they thought was worthy of being recorded or noted down. Yet we (and they) discovered that they had an awful lot to say to DUCHESS.

As an example, people were willing to say that they did not think that they had the right training to do the job that they were in but they did not want to tell anyone because the person before them had managed it. They thought that if they raised this with a human interviewer, it would reflect badly on them as being in some way incompetent or less skilled than the person that came before them. So they would not have raised this fact with a human interviewer but were happy to tell DUCHESS.

On the other hand, our hypothesis that a DUCHESS interview would not be as good as a human at generating rich insight was shown to be incorrect. In fact, we found that being obviously a machine and not a human gave people the space to talk whilst trusting that what they say is not going to be attributed to them. So in cases where the interview subject matter was potentially controversial, DUCHESS was better than a human interviewer just because it removed the inhibition that the fear of judgement caused.

The most important thing we discovered in the initial testing was that peoples' honesty led to them raising issues we had never thought to ask about. For instance, people talked about bullying and harassment even though we asked no specific questions about either. Just giving them the opportunity to talk meant they were willing to share more than we asked.

Finally, we discovered that just the act of doing the interview seemed to benefit some people. People leaving the room after the interview said they found it cathartic; some said they needed a good cry after their interview. It was almost therapeutic, just to be able to sit in front of a laptop and get everything off their chest, knowing it was anonymous. Of course, some people would rather talk to a human, but in this case, for post-deployment interviews, there were insufficient human resources available to conduct in-person interviews. Interestingly when we asked whether people would rather a human or DUCHESS, some said human, some said DUCHESS. However, if we said a human is not an option, so the choice was a survey or a DUCHESS interview, all asked went for DUCHESS.

At this early stage the only niggle was how comfortable people were in talking to AI and how easily it picked up what they said. For some people, it took a question or two to warm up, and sometimes the quality of the microphones was poor, but these issues did not seem to detract from the opportunity of being able to talk.

Development

Having completed the initial proof of concept, we got a second and then third round of funding, which took us up to 2020. At that point, we had a fully functioning application that could be installed on a network and did not rely on any internet connectivity. It could be used by any number of people within an organisation. Indeed, a particular organisation within the MOD asked us to install DUCHESS on one of their closed networks.

However, when Covid-19 hit, we were unable to show the application to anyone or install it on those closed networks given we could not meet in person. So we invested in a cloud-hosted version. This would allow us to send anyone, anywhere, a link to the web app, and they could do an interview online. This took us away from our niche use case, which was to be able to do interviews on a range of subjects, to multiple use cases. And it took us away from the military, who needed to run it offline. But we felt that if we were going to exploit the technology we had developed, then we had to make it more accessible to other industries. So our company funded the development of the web app internally.

The web app version was Azure Cloud hosted, so we were able to use Microsoft's data protection services. Our privacy policy can be accessed from the application and details to interviewees how we will use their data, so they are assured that it will not be shared and that, unless they are told otherwise, their transcript will not be read by anyone within their organisation. At the start of their interview, they are advised not to talk about any information that would go above the security classification of the system that they are using DUCHESS on, although there is no way to stop them from doing so.

There are two ways that DUCHESS can be used. One is in a fully anonymous way, where the same link is sent to every user, who clicks on the link without needing to log on, and they do their interview. We collect no data from the user or the device that they are conducting their interview on. The other way requires users to log on and we collect their personal details, which they are told upfront in the email containing the link; alternatively, if the app is hosted on an intranet, they are told in the introduction. It is made clear to them whether this interview will be anonymous or whether it is possible they might be identified. In some cases at the end of their anonymous DUCHESS interview we have linked to a survey to collect some demographic data, but they can choose what they share.

The underlying DUCHESS technology has also evolved, particularly with the advances that have been made in LLMs. In 2024 we invested in the development of DUCHESS version 2.0, which gave a step-change in the conversational nature of the experience through the use of commercial LLMs. Our initial follow-on questions could prompt people into talking more about things that were important to them. But DUCHESS version 1.0 did not have a deep understanding of the context people were talking about. It could determine that this person felt strongly about this 'thing' but not what this thing was apart from being a word or a short phrase. The more people interact with AI, the higher their expectations become, and so to keep pace with the latest technology and continue to be innovative, we chose to make use of LLMs' to enhance the experience of holding a conversation.

One key feature about DUCHESS is that it does not need any training. In terms of the interviewee completing an interview, it is designed to be purposefully simple, with very few buttons. Users can choose to watch an explainer video before their interview, but the idea is to be very much click and play: the user clicks the link, talks to it, ends the interview and it's done, from an interviewee perspective. Clients send us their ideas for questions, which we adapt slightly to be in a form that DUCHESS will work well with, so they are well-formed, open questions that encourage users to talk without leading them. It's up to the client what to do with the results. In some cases, we have written an analysis that summarises the insights that can be gained from across the interviews. In other cases, clients have analysis capability within their organisations, and so we pass the data over to them and they write their own report.

Adoption

In seeking funding for an innovation, as we did in the early stages of developing DUCHESS, highlighting the breadth of usage and impact means the money spent is more likely to generate a good return on investment. Everyone in the military that we spoke to, and demonstrated DUCHESS to, loved it. They would often come up with new use cases relevant to their area or part of the organisation and then ask how they could get access to it. That proved to be a good question to which no one seemed to know the answer. Of course, some people within MOD knew how to deal with it technically but none knew what the commercial process was.

The main question that we needed someone in the MOD to address was who owns an application that cuts across the organisation. Departments that wanted to use it could not fund its implementation on their systems because they could not fund something that others would benefit from. That is not a unique situation within defence; there are plenty of shared capabilities but these tend to be major defence platforms for which senior military officials reach some agreement. Something like the DUCHESS AI, which is a 'business as usual' or back-office capability, falls uncomfortably between two stools. It is not a 'per person' application like Microsoft Word, Excel and PowerPoint, where the cost can be amortised across the organisation, nor is it a major defence platform which would warrant agreement amongst the seniors. With the offline DUCHESS version 1.0 we were using, we were pointed to departments such as Defence Digital and the Defence AI Centre, which were touted as the backbones of the system for such AI applications. But we soon found that they had so much demand that they wanted someone to stick their head above the parapet as the owner and funder. Alas our rather depressing observation was that, as DUCHESS is cross-cutting, everyone wanted to use it but no one wanted to own it. And without an owner there was no funding, no commercial section to do the contracting for it, and no management of it. We could thus not progress because the MOD wanted it to sit on their systems for data protection reasons.

The solution that we hit upon (by accident) was to ditch the focus on defence-specific requirements. Having started with an offline system to be used on classified machines, we developed the cloud-based application which had the happy effect of allowing higher-performance transcription services. We also gave up on various use cases that had been suggested and focussed on the original lessons-learned interviews, for which we some expertise in terms of the 'seedling' open questions. After that, irony of ironies, we got our first defence client – the Joint Analysis and Lessons Learned Centre (JALLC) of the North Atlantic Treaty Organization (NATO).

Interestingly, NATO JALLC used DUCHESS to conduct a lessons-learned study on something common to nearly all organisations – people's experience of working from home during the Covid lockdowns. Security was not an issue as the subject matter did not relate to operations, so the online system was perfectly suitable. The NATO JALLC team even published a paper on the experiences of using DUCHESS.[2] Although they used version 1.0, it provided a good example of how DUCHESS could reduce the effort required to capture lessons (from more than 130 hours to less than 5 hours). Soon after, a team within the Canadian military contracted us to do something similar, then again for a different study, then again. Then, at last, a department within the RN (the original sponsors of DUCHESS) funded an implementation on their systems to allow classified interviews. We now had momentum, with clients for both the cloud and offline versions which gave other defence organisations the confidence and motivation to use it. Most importantly, we learned our own lessons on how to get it adopted. DUCHESS is now implemented on the systems of several militaries and a small portfolio of defence industry clients use it for use cases as varied as wargame feedback and cultural change initiatives.

Reflections

Apart from the technical lessons that we managed to incorporate as DUCHESS evolved, we learned an awful lot about software development. We underestimated how much effort was required to take something from proof of concept, which we had done a lot of in the past, to something that is robust and simple enough to put into the hands of a user. This is particularly the case because potential clients and users saw themselves as paying for a service, or

---

2 NATO Communications and Information Agency, "NATO's Joint Analysis and Lessons Learned Centre: On the Cutting Edge of Innovation," NATO, 3 May 2023. Accessed 2 August 2025. https://www.act.nato.int/article/natos-joint-analysis-and-lessons-learned-centre-on-the-cutting-edge-of-innovation/.

at least for 'software as a service'. We never offered it as such but that seemed to be the preferred starting assumption. So it was not enough to demonstrate how well the AI algorithms performed; we also had to demonstrate ease of use. This may be blindingly obvious to software development firms, but AI developers do not always come from that background. A military officer seeking to set the requirements and procurement of such a system must always consider as many different ways of getting the benefit (starting with software versus the service versus 'software as a service'), and then include requirements for the ease and robustness of use of the application as much as the performance of the AI itself.

DUCHESS also highlighted the trade-off between 'the best' and 'the easiest'. We had much experience with proof-of-concept work where the key objective was demonstrating the performance of the algorithms. Naturally, we then wanted the customer who first sponsored the work to get the most functionality and performance, but to do that meant overcoming many hurdles related to organisational and information technology issues. When we switched to the online system, we made it easier to use – not for the original sponsors, alas, but to others for whom many of the original requirements were of less (or no) importance. That built experience and reputation, which in turn drove a desire to overcome the original difficulties.

For a military officer involved in the requirements and procurement, funding research and development into a capability and then accepting that someone else benefits first is hardly career enhancing. However, perhaps it might help to update the old idea of 'incremental acquisition' and apply it in the context of a holistic view of users and use cases. It might also help to consider a full range of use cases to see which can be met by the smallest number of requirements, in order to demonstrate early benefits to the military enterprise and provide a means of getting feedback to improve the next increment in capability.

Our DUCHESS experience also illustrated the need to incorporate advances in the underlying technology. Again, for a software firm this is a no-brainer. For an AI firm wedded to its algorithms, this can be more challenging. But it can be essential, especially as things such as security protocols are always evolving. For instance, long after deploying DUCHESS using Microsoft Azure, it continues to require a huge amount of work to keep it functioning on that platform – which we need to factor into our architecture decisions, particularly how much we can rely on things that we have no control over.

We also have to factor in what is required to future proof the application. The DUCHESS interview engine links to a database, and that link is provided by a function within the Azure platform, and elements of that function will expire so we will need to rewrite it. There are many links as DUCHESS goes back and forth, with questions and answers, writing data. It becomes a significant part of the application, therefore a significant investment to keep up with the latest acceptable versions. While this might seem like an issue that the AI company should or could deal with, we suggest that a military officer might wish to include the need to update certain aspects in the requirement. The challenge here is how to price that, given future technology advancements are unknown. However, having something akin to a fixed budget for upgrades at regular intervals combined with practices such as MoSCoW prioritisation (Must have, Should have, Could have, Won't have) would go a long way to limiting the frequency of painful replacements of capability.

The final reflection about DUCHESS concerns the impact on people. The reflex view is that AI will replace people. This is partly the result of automatically grouping AI with automation. The fact is the machines can automate things, and hence remove people from a process, without AI. Obviously, AI can automate different things and replace different people. However, DUCHESS is one of the purest examples we have seen where the AI is specifically designed to make the best of what people know and have experience of. Almost every large organisation has people, or teams of people, tasked with improving processes and learning lessons. Every large organisation also has lots of people with experiences and insights that it could learn from. In our experience, no large organisations capture the insights of all of their people. Instead people are prioritised or sampled, which leaves potentially many experiences and insights untouched. DUCHESS is one example of an AI technology that can leverage up the capacity of those seeking to capture lessons, so the capacity of people to turn experience into insights is matched by the capacity of the organisation to capture and learn from them. Add to that the hierarchical and cultural issues that separate those in management seeking to learn from those at the shop floor or front line, and DUCHESS shows how culture and judgement-free AI could transform organisational learning and, ultimately, transform improvement from within.

5

MALFIE: AI That Explains and Prioritises the Outputs of Other AI

The complexity of AI systems is a double-edged sword, wherein enhanced capability is paradoxically paired with decreased explainability. – 'Mad Scientist', US Army Training and Doctrine Command community of action[1]

Key Takeaways

DUCHESS was artificial intelligence (AI) that captured the lessons that humans had learned; MALFIE (Machine Learning Fuzzy-Logic Integration for Explainability), however, is AI that learns what other AI has learned and explains it to humans. Specifically, it monitors a number of AI applications that are looking for anomalies in international shipping and learns how to explain to an operator why these other AI applications have picked certain ships out as anomalous. It turns the need for explainability of increasingly advanced AI outputs from the paradox, highlighted by those such as the 'Mad Scientist' community of action quoted above, into an example of a feature that AI should have as a matter of course. However, it also highlights that 'explainable AI' is not a simple matter of applying a few data science techniques that some AI enthusiasts use to wave away concerns.

MALFIE is also a counterpoint to DUCHESS in that much of the preparation phase of the utilisation staircase discussed in Chapter 1 was

1   Mad Scientist Laboratory, "11 Artificial Intelligence (AI) Trends," 14 December 2017. Accessed 25 November 2024. http://madsciblog.tradoc.army.mil/11-artificial-intelligence-ai-trends/.

addressed very early with (or in some cases by) the potential military user. The potential users had already tested other AI so it was possible to quickly address questions of what existing (human) process it addresses or is similar to; how it is different from other AI; what it does better or allows one to do that was not possible before; what perceived risks it deals with and whether they are relevant; how the military user would actually use it; what risks it introduces; and so on. As a result of this well-informed user-pull, even the normally fraught question of whether there is enough time to mature and integrate the AI was addressed relatively quickly after phase 1. That left only three questions of the utilisation staircase for the team developing the AI to address.

The first key takeaway relates to this question from the utilisation staircase from Chapter 1: *Is it as complex as needed (as simple as possible) or as complex as possible?* The MALFIE case shows that simple, older techniques (which may be looked down on by academics and technologists) can still meet the definition of AI and deliver something of use. In this case the 'task normally conducted by a human' was indicating which among many potentially suspicious movements at sea were priorities for further investigation and what exactly made them of interest. The basic 'anomaly detection' was conducted by several advanced AI systems whilst MALFIE provided an overarching priority and explanation covering all the other AIs. Hence an operator only had to deal with MALFIE rather than monitor several different AIs simultaneously.

The second takeaway relates to the question: *What will be done to address the new risks?* MALFIE showed that an AI 'overlay' can be used to overcome some of the traditional concerns about the use of AI, specifically the challenge of managing a plethora of AI systems and outputs and needing to understand what the AI is telling the responsible and accountable human. MALFIE monitors the inputs and outputs of multiple AI applications, learns how these relate to each other, and converts that learning into language that a stressed military operator can act on (as opposed to being a quantitative 'explanation' that only a PhD data scientist can use). This is in many ways closer to what humans do in practice than the theory of rational decision-making would suggest, especially in the case of rapid decision making in stressful situations.

The final takeaway related to the question: *What is the level of ease of use?* MALFIE showed how the seemingly simple and fundamental idea of requiring the AI to provide an explanation, which is so often raised by those concerned by or suspicious of AI, can end up being a practical problem. It is not necessary for someone to give a full explanation to every question asked. Reasoning and explanation constitute only an intermittent event in real-life human interactions. The same is true with AI-human interactions. This realisation led to two other AI applications, namely a Multi-Agent Dialogue module (MADM) and Red Mirror, needed to understand when explanations were timely and relevant given the situation and context, respectively.

Context

"Get the machine that goes 'ping'." "And get the most expensive machine in case the administrator comes!" – Monty Python's The Meaning of Life

In the 1983 Monty Python film *The Meaning of Life*, there is the lovely 'Machine that goes ping' sketch. Two surgeons are in an empty operating theatre. They then order the theatre staff to bring in the machines. Only after the theatre has been filled with various machines, much to the delight of the surgeons, does someone realise that they are missing the patient. After some searching a pregnant lady is spotted waiting behind a bank of machines and placed on the theatre table. The administrator arrives, expresses his pleasure that they are using the machine that goes ping and, when told they will be performing a birth, comments "amazing what we can do these days". The labour proceeds, with the noises and pinging of the various machines around the woman (none of which are actually used by the theatre staff), the baby is born and the theatre then cleared.

A young ex-naval officer who joined our team specifically to work on MALFIE described something very similar in regard to a reporting and management system that had been integrated onto a ship on which he had served. Although it had, by all accounts, excellent performance in terms of spotting and reporting issues, in practice it produced so many alerts and messages that crews simply turned it off. While each individual report and management function on the ship had its merits, integrating all of them just led to a cacophony of pings, beeps, buzzes and whistles which often led to them being ignored or switched off.

One of the most widespread use cases for AI in the 2010–2020 timeframe was anomaly detection. Anomalies are common across all industries, and AI could make good use of all the data generated, collected and shared across the burgeoning networks. Many use cases for detecting anomalies required only a few types of input, and both supervised and unsupervised learning techniques could be used. The output was conceptually simple – just alert a human operator to the anomaly. That is useful if the system being monitored is in a fairly steady state, such as an industrial process or security perimeter, and the anomaly is rare and thus can trigger a relatively standard response. Unfortunately, one of the characteristics of defence, and particularly military operations, is that there is often no steady state. Variability is a key facet of the military, with each side seeking to gain an advantage over the other, and each side then reacting whenever an action is taken.

Also, the range of contexts, situations and players in a defence scenario means there is a greater range of 'normal' behaviours, which then means a greater range of potential anomalies to consider. Finally, the scope of some defence activities, such as satellite surveillance of sea lanes and air lanes, means that there could potentially be hundreds, or even thousands, of anomalies detected every second. All this means that there is a great risk of any AI system simply overwhelming a human operator or overseer with alerts. That in turn risks overwhelming the capacity of the overall process to deal with those anomalies until the machine that (endlessly) goes ping is turned off.

There are two solutions. One is to fully automate the entire process from detection to execution, but that is a massive undertaking for all but the lowest level of tactical decision (for example, automated protection systems such as Phalanx and Goalkeeper on ships). The other solution, which definitely requires humans to be part of the process, is to use a prioritisation system. However, in many cases the criteria for prioritisation may not be easily defined. That is where explanation becomes useful, as it can be a richer and more context-sensitive way of providing evidence to support the prioritisation of a human operator's or overseer's attention and subsequent action. Of course, the benefit of using the AI only accrues if the act of providing and reviewing the evidence and priorities does not slow down the end-to-end process to such an extent that it becomes slower than the purely manual process, or too late to do anything.

Inspiration

The idea for MALFIE came at a hackathon which had become a popular way for organisations to generate and test ideas. The challenge put forward was how to use the information from the Automatic Identification System (AIS) on ships to detect anomalous behaviours. AIS is only a requirement for ships above a certain size, and ships seeking to avoid detection could simply turn off their AIS transponders, so this was clearly an initial investigation. Interestingly, a trial had already been conducted with a commercial anomaly detection system. However, because it had been developed to spot anomalies in air lanes, it did not prove to be very useful in the maritime domain.

DIEM's Sarah Vincent-Major attended the hackathon and, with her background in the Maritime Warfare Centre (MWC) and Navy Command Headquarters plus her involvement with DUCHESS, quickly hit upon a key issue with the challenge as presented. Anomaly detection using AIS (or any other sensor data, for that matter) was conceptually straightforward, and there were many technological options that could be taken forward. What mattered, however, was presenting the outputs of the anomaly detection algorithm or algorithms in a way that was actionable for an operator. Sarah brought together a team from the various hackathon attendees, which included individuals from a software firm, a defence prime and a couple of academics. This team, which they named '404 Page Not Found' in recognition of some problems with access to data, created a simple mock-up of the concept followed by a simple demonstration using the sample AIS provided at the event.

Team 404 did not know, at the time, about the existing anomaly detection system that had been trialled as part of a regular military exercise. The system had been developed by an Australian software company as a tool to establish aircraft 'patterns of life' (essentially within normal air lanes), and hence to spot anomalous air tracks. In this context 'anomalous' usually meant aircraft that were not conforming to air lanes. A version adapted for the maritime surface domain had been acquired and, as part of an exercise, it was fed with live AIS and radar data. This highlighted several issues with the state-of-the-art anomaly detection as it was then. First, sea lanes are less well defined than air lanes, so the system produced far more alerts of anomalous ships than should be dealt with. Second, whilst the system could indicate the reason why a vessel was not an anomaly (largely because it was on a known route), it did not provide reasons for why the vessel was an anomaly. Just saying 'the vessel is not on a known route' (the opposite of 'not an anomaly') is not sufficient for an operator to prioritise that contact over many others which have the same 'reason', or to justify the need for further investigation. Third, the system only assessed a vessel's current position, speed and direction to determine whether it was an anomaly. It did not spot vessels that were suspicious because of their previous behaviour. Finally, there was no ability to 'rewind' the system's visualisation to assess a vessel's past behaviour or record past behaviour as evidence.

The company that had produced the system had subsequently ceased trading and so it was not possible to develop it further. The hackathon, therefore, was an attempt to explore ways to overcome these observed limitations in a new system. Team 404 identified that the challenge in the surface warfare domain was not in detecting anomalies per se, as systems had already demonstrated that this could be done. The challenge was in explaining why the detected anomalies were in fact anomalies in order to give military decision makers the understanding and confidence to take the appropriate action. If there could be an overlay to the anomaly detection system, it could solve the problem of too many alerts and no explanation, rather than replacing the system itself. This idea, which might have been considered non-compliant or out of scope under a normal procurement process, received very positive feedback from the hackathon organisers. DIEM were asked to submit a full proposal as part of the competition organised by the Ministry of Defence's Defence and Security Accelerator (DASA).

Inception

The other three case studies in this book put forward an idea that fit within a general area that defence was interested in applying AI. In the case of DUCHESS, DIEM came up with the idea of the AI interviewer and then submitted it as a proposal to a DASA competition related to the general area of 'defence people'. We identified the problem that DUCHESS might help with, and a sponsor was then found who had a particular interest in that problem and who could provide input throughout the various phases. The benefit of this 'idea push' was that we had the freedom to explore variations of the problem as the application developed and we could see what it was best it.

MALFIE was the opposite. The potential user had a specific problem, and we proposed MALFIE as the solution. We had to suggest something that the user could see was indeed the solution (or at least one solution). Our particular skill in terms of the backend AI algorithms would not be enough; the whole problem focussed on explaining and prioritising the solution for a stressed military operator so it would be particularly ironic if we demonstrated an algorithm that users could not easily understand. From the start we needed a bigger team than we would normally have in order to cover both the back-end algorithms and a front end that provided a good user experience, and ensure both were properly integrated along with the potential input data.

With Sarah and me as leads, the team consisted of two people from a small software company who were completing part-time degrees, a professor from the university where they were studying, and two people from a Ministry of Defence contractor who had experience of developing user interfaces for the Royal Navy. Putting the bid together was relatively straightforward and once we were informed we had won, contracting went quickly, although there were some issues with a small company subcontracting a big company and an academic institution, largely driven by the different pace of work.

At the kick-off meeting, we received interesting technical feedback on our proposal. Some technical experts felt we were "lucky to get funded" because our proposed techniques were "quite old". We could not argue with this observation as we were proposing to apply an 'ensemble' of fuzzy logic and machine learning (ML) – hence the name 'MALFIE', which stood for machine-learning and fuzzy-logic integration for explainability. Both these had been around for decades, but we chose them because the stakeholder had large amounts of unlabelled AIS data but no labelled data with which to train a system. It did, however, have operators who could advise what to look for and how to prioritise multiple anomalies. So our approach was to use ML techniques to automate the process of generating fuzzy logic rules which could then be assessed by operators and compared to actual events. The idea of automating the creation of the rulesets was innovative. However, the combination of two old techniques, even if done in a novel way to achieve something that had not been achieved before, was less interesting to the technical experts than using newer techniques such as convolutional neural nets (CNN). Ironically, CNNs were among the technologies being applied to identify anomalies in exactly the way that organisations had found could not be acted upon because of the lack of prioritisation and explanation.

The military members of the marking panel, however, took the opposite view. For them, our proposal was addressing a problem that had not been addressed before. The novelty of the technique was largely irrelevant, although we subsequently found out that there was some hope that the use of older techniques would make things less expensive and quicker to get into service, compared to the CNNs already being applied to other areas and taking significant time and effort to mature for use by military operators.

This divergence highlights something seen again and again in defence: the tension between the focus on practical solutions now versus new technology for the future. Those who commission defence research tend to shift from one side to the other every three to five years depending on the defence context. When there is an ongoing campaign such as the deployments in Iraq and Afghanistan, the focus is on 'quick wins' that the military can rapidly exploit. Once those campaigns are over, there is a swing back to looking at 'generation after next' technologies. The rise in interest in AI is unusual in that both views could apply depending on the stakeholder. The military naturally see it as something that could have a rapid impact, having seen the way technology such as ChatGPT has done so, whilst the technology and research community (having seen how that same technology has advanced) sees it as something that could be revolutionary and transformative. Both are potentially true, but as budgets are always constrained someone must choose between these two opportunities.

The dissension regarding the decision to fund our proposal proved useful as it led to us defining three hypotheses that we could seek to prove. That helped us structure the work in a way that aligned with the scientific method. Our first hypothesis was that the raw output of an anomaly detection algorithm, usually given in terms of a probability or index, is not a sufficient explanation to be able to justify a specific action compared to a natural-language ruleset. The second hypothesis was that fuzzy-logic rulesets that are useful to provide explanations can be generated automatically using unsupervised ML rather than using the input of subject matter experts (SMEs). The third hypothesis was that the natural language input parameters required for the rulesets can also be generated automatically using unsupervised ML rather than by the input of SMEs.

Technology

Our proposal reflected a combination of what had been presented at the hackathon and what we had learned from other research projects related to maritime command and control. We sought to bring together technology that covered three levels of functionality (see Figure 5.1).

First, we had to find anomaly detection algorithms. These had to use techniques that could be applied to a large amount of unlabelled AIS data from which a wide range of normal and not normal (anomalous) behaviours and patterns could be spotted. This was effectively just providing equivalents

Presents the alerts and priorities and allows drill-down into explanation

Prioritises tracks & explains why they are of interest
- The more anomalies looked for, the more likely a track will be anomalous
- Analyses the combination of anomalies and patterns in order to explain*

Detects a range of different anomalous behaviours or patterns of interest [could be any type of anomaly detection algorithms]

Application
- Control panel and chart
- Vessel details
- Trends
- User-defined areas

Machine learning of ruleset to explain

Clustering to prioritise anomaly pattern

Input, training & output integration

Anomalies of interest	Patterns of interest
• Movement criteria	• Collisions
• Vessel criteria	• Formations
• Time criteria	• Patterns

*e.g. Police will stop a vehicle being driven "too carefully" because this behaviour is very unusual and may indicate the driver is trying to hid something

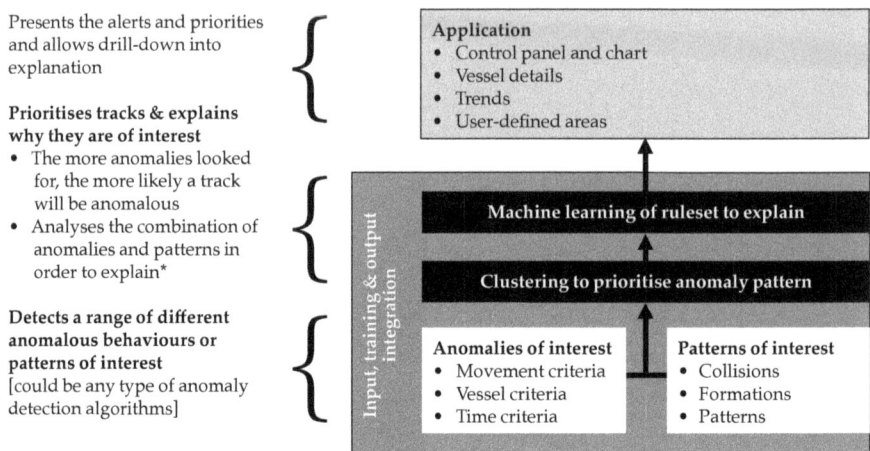

Figure 5.1: MALFIE levels of automation

of the anomaly detection system already tested, but over which we had sufficient control to link to our explanation and prioritisation concepts. As we carried out a literature review of potential algorithms and discussions with stakeholders and potential users, we realised that the term 'anomaly detection' (for our purposes, vessels doing something that is not normal) only covered part of the problem space. The other part was the detection of vessels exhibiting a pattern of interest, irrespective of whether it is normal or not. As the MALFIE concept needed to be able to explain the outputs of any underlying detection system, we decided to have both an anomaly detection system and a pattern of interest detection system.

For the anomaly detection slot we chose the Traffic Route Extraction and Anomaly Detection (TREAD) algorithm[2] developed by the NATO Centre for Maritime Research and Experimentation. This algorithm calculated three metrics for 'normality': the probability of being on a known route based on the routes taken by other vessels in that area, the likelihood of following a known route based on the vessel's previous routes and the likelihood of staying on a known route based on a projection of where the vessel should be versus where it currently is. These three metrics combined gave an overall score for normality, which is the inverse of how much of an anomaly it is. The beauty of TREAD was that it allowed these metrics to be calculated

---

2   G Pallotta, M Vespe and K Bryan, "Vessel Pattern Knowledge Discovery from AIS Data: A Framework for Anomaly Detection and Route Prediction", *Entropy* (2013). https://doi. org/10.3390/e15062218.

in different contexts such as for specific areas, seasons, types of vessel or periods of time. Hence, TREAD could determine whether a fishing vessel, for instance, is anomalous compared to all vessels or other fishing vessels, or whether it is anomalous compared to what it did in the same season last year.

For the pattern detection we developed a bespoke Collision and Formation Pattern (CFP) algorithm. This identified when individual or pairs of vessels were on a collision course, in formation, rendezvousing, mothershipping, or circling or converging. Each of these patterns was highlighted by military SMEs as something that they would like to be notified of, as each could indicate a priority to investigate. Whenever the algorithm detected that a pair of ships were exhibiting any of these patterns, it would output the distance between them (where the smaller the distance, the greater the level of interest).

With the underlying anomaly and pattern detection algorithms in place, the second level of functionality was the ability to pick out the key parameters to base the explanation on. This is a major difference between 'explainable AI' for military operators versus data scientists or AI practitioners. For data scientists, techniques such as SHAP (SHapley Additive exPlanations) and LIME (Local Interpretable Model-agnostic Explanations) provide quantitative comparisons of the contribution of different inputs to the output, or the sensitivity of the AI outputs to the different inputs, which are useful for understanding the behaviour of the AI overall. However, the type of explanation needed to support military decision making relates to the specific output at that point in time, and this requires narrowing down what is presented and how.

To illustrate this, we looked closely at the application of AI to support the air defence process within a ship's operations room.[3] We found that the type of explanation an air warfare officer (AWO) might give to a commanding officer as to why, for example, they wished to send a level 3 warning to an approaching aircraft could be as simple as "I am concerned by this contact because of its altitude, speed and heading". We identified 40 input parameters that the AWO and their team would consider to determine the radar track and how to respond, but saw that the team would focus on the small number of things that they thought were most significant at that moment. They were not ignoring the other 37 parameters; they had learned that highlighting only the key ones allowed for a quick decision (speed in making the correct

3 Richard Scott, "Machine speed warfare: UK tests naval AI decision aids in ASD/FS-21 exercise", *Jane's International Defence Review* (July 2021).

decision is vital in these situations). We also noted that they did not always indicate the scale or value of the parameters. Again, this was context specific as everyone could see the altitude, speed and heading on their screens for themselves if they needed that information. They also know, through training together and the processes they follow, that others would understand that they meant "low altitude, high speed and heading straight towards us" in this context. Hence adding in the precise altitude in metres and speed in knots would slow things unnecessarily. By contrast, if the same officer was asked to explain their decision at a board of enquiry, they would more likely be very specific about these parameters and may even add in a few more, but would still be unlikely to reel off the value of all 40 parameters.

From our previous work we knew that fuzzy logic provided a good mimic of how operators in the maritime area structured the inputs to decisions, but we needed to automate the process of generating the fuzzy rules from the outputs of the anomaly detection algorithms. This required a three-step process, each with its own set of ML techniques. The first process was to normalise and then categorise the different underlying anomaly detection outputs so that they could be considered together (equivalent to the definitions of high and low in the example above). In the example of the TREAD 'probability of being on a known route' indicator versus the CFP 'formation' indicator, TREAD takes values between 0 and 1, whilst the CFP could take any distance, so this must be normalised. Moreover, it may be that most vessels get a value of 0.8 to 1 for the TREAD metric with some vessels getting scores of 0.4 to 0.8 and a very small number getting less than 0.4. This would translate to low, medium and high indicators of being anomalous. For the formation distances from CFP, the values that relate to far, medium and close would have a different spread. To automatically define the categories, or sets of values, we used a bespoke variation of the kernel density estimation (KDE) algorithm,[4] which we called KDE Set Definition (KDE-SD), to machine-learn the most appropriate way to define the categories within each anomaly or pattern detection algorithm's output metrics.

The second process was to determine whether the anomaly or pattern-of-interest score from the underlying anomaly detection algorithms fell into a category or level that made it truly an anomaly from an operational point of view ('I am concerned by this' in the example above). This might seem like a repetition of the anomaly and pattern scores, and it would be if the

---

4   KDE is effectively a one-dimension clustering algorithm.

point was to judge the level of anomaly on a single indicator. However, the more indicators of anomaly and patterns examined, the greater the chance of something looking like an anomaly. So, if each vessel is assessed against 100 different indicators, the chance that one will generate a high anomaly score will be itself high. Hence, if there are many vessels, they will all look anomalous in some way, resulting in a 'machine that goes ping' and an operator unable to decide which is truly the priority. In this case, the profile of anomalies across all the indicators becomes important, rather than the individual anomalies (or even an average).

As one example, using the TREAD metrics of following known routes, all fishing vessels would seem anomalous, because they follow the fish rather than the sea lanes followed by the other ships. A fishing vessel taking the same route every day would be of most interest because it might indicate some nefarious activity such as plundering a wreck. As another example, when the police see someone driving very carefully and 'too perfectly', they might pull them over to see if the driver is being cautious because they are trying to avoid doing anything that would get them breathalysed.

We tested several clustering algorithms (k-means and DBSCAN being the most well known) to create groups of similar anomaly and pattern score profiles. In Figure 5.1, these two steps of the process are grouped under clustering.

The third step of the process was to identify which input parameters are needed to provide a good explanation, the equivalent of deciding to highlight altitude, speed and heading in the example above. To keep in line with operational process and ensure speed, we tested two decision-tree algorithms (C4.5 and CART).

The third level of functionality was the combined visual and verbal (in the form of text) interface that presented the current situation and allowed exploration of particular vessels and the explanation of their anomalies, patterns and priorities. This was the subject of a second literature review to identify examples of visualisation of maritime anomalies and their explanation, and from these develop design rules that could be followed. Our analysis of the characteristics of the classic example of Minard's visualisation of Napoleon's Russia campaign as a visual explanation (or infographic) as well as recent maritime situational awareness applications led to several design rules that we applied. A key part of this was defining useful graphical techniques, of which one that stood out was the use of parallel coordinate plots to show comparisons against multiple parameters.

Alongside the visualisation was the use of natural language processing (NLP) technology to generate the wording for the description and explanation of the anomaly and its priority in a way the operator can readily understand and act on if needed. This is the same technology deployed in DUCHESS but in the opposite direction, that is to present information rather than ask questions to gather information. Obviously, the NLP technology which was leading edge in the 2017 to 2019 timeframe when we worked on DUCHESS and MALFIE subsequently led to the large language model technology that is now so popular.

To progress on this front, we had to delve into the theory and philosophy of explanation. The need for 'explanation' with respect to AI is thrown around quite readily without much understanding of what it means at a technical level. The starting point was the Oxford English Dictionary definition of 'explanation':[5]

> A statement or account that makes something clear: *the birth rate is central to any explanation of population trends.*
> • a reason or justification given for an action or belief: *Freud tried to make sex the explanation for everything.*

This definition refers to three styles of explanation: causality (births cause population), reason and justification.

Aristotle's theory of causation[6] was the first attempt at defining how to answer the question 'why' and consists of four causes. First is the material aspect of the change caused: for example, that vessel is anomalous because it appears to be on dry land. Second is the form or formal arrangement, shape or appearance of the thing changing or moving: for example, that vessel is anomalous because of the shape of its route. Third is the agent of change: for example, that vessel is anomalous because it has suddenly increased speed. Fourth is the end or purpose: for example, that vessel is anomalous because it is a ferry but off the ferry route. Many of the subsequent theories of explanation and reasoning can be traced back to these four cases. Teleology is a popular approach for explaining ethics and is rooted in the 'final purpose' cause.

---

5 Oxford Dictionaries, "Explanation". Accessed 17 August 2025. https://www.oed.com/dictionary/explanation_n?tab=factsheet#5067609.
6 Andrea Falcon, "Aristotle on Causality," Stanford Encyclopedia of Philosophy, last modified 7 March 2023. Accessed 25 November 2024. https://plato.stanford.edu/entries/aristotle-causality/.

There appears to be no standard description of a reason or justification (the Oxford English Dictionary definition calls it a form of explanation, thus creating a circular definition). There are, however, different types of reasoning, which can be subtle. Deduction is where a logically certain conclusion can be reached if all the specified premises (or observations) on which it depends are true: for example, if the transmissions have the correct code for HMS *Dragon*, there is visual identification that the silhouette is of a T45, and it has 'D35' on the side, then it is a friendly ship. Induction is where specified premises contribute some evidence to the truth of a conclusion: for example, the vessel has its AIS on, it is on a ferry route and travelling at 10 knots, so it is probably a ferry. Finally there is abduction, where unspecified premises are used to derive a conclusion: for example, the vessel has turned its AIS off, it is not on a known route, its speed is higher than normal, it is a suspicious vessel.

These types of reasons form the logical basis of an argument, which is, technically, different from an explanation. Arguments show that something was, is, will be or should be the case, whereas explanations show why or how something is or will be. Nevertheless, from these definitions of reasoning, we surmised that a reason could simply be a statement of some or all of the premises used to reach a conclusion: for example, 'I didn't fire on it as the transmissions suggested it was HMS *Dragon* and it had D35 on the side'. By contrast, a justification involves stating the conclusion arrived at: for example, 'I didn't fire on it as it was a friendly ship'.

Carl Hempel[7] defines an explanation as "an argument to the effect that the phenomenon to be explained … was to be expected in virtue of certain explanatory facts." His theory of explanation[8] suggests that an explanation consists of two types of statements: initial conditions (C) and 'law-like' generalisations (L). These combine to cause or explain an event (E). Hempel described two types of explanation, which mirror two of the three types of reasoning. Deductive-nomological (DN) is where the laws are universal generalisations: for example, the vessel is transmitting UK military codes (C), vessels that are transmitting UK military codes are to be marked as friendly (L), the vessel is marked as friendly (E). Inductive-statistical (IS) is where the laws have the form of statistical generalisations: for example, the vessel is

---

7 C G Hempel, *Aspects of Scientific Explanation and Other Essays in the Philosophy of Science* (New York: Free Press, 1965).
8 Internet Encyclopedia of Philosophy, "Explanation". Accessed 25 November 2024. https://www.iep.utm.edu/explanat/.

not conforming to a known route (C), most vessels that are not conforming to a known route are suspicious (L), the vessel is marked as suspicious (E).

There are two fundamental criticisms of Hempel's theory. One is that in the IS case, the probabilities implicit in the law need to be high for it to be a useful explanation. The other is that it is possible to link conditions to events through unrealistic causality. For instance, the vessel is not conforming to a known route (C), fishing vessels often do not conform to any known routes (L), the vessel should be marked as 'fishing' (E). The fact that the vessel was not on the known route as a reason for marking it 'fishing' is technically an explanation, based on these statements, but would not be considered realistically causal.

Bas van Fraassen[9] defends 'anti-realism' through what he termed 'constructive empiricism'. He characterised explanation as an answer to a question of 'why … x rather than y?' where the different options may be implicit or explicit. In our case this would translate into 'why is this vessel more likely to be anomalous rather than that vessel?'. In specifying the alternatives, the context and hence relevant realities can be more easily identified, and the probabilities in the laws do not have to be high per se, but need to be higher for one option than the other.

After considering these theories we settled on three styles of explanation. The first was the explicit explanation, containing a statement of the condition, such as a parameter and its value that has triggered the relevant law: for example, two contacts in manoeuvring in formation. The second was the implicit explanation, consisting of a reference to the condition, such as just stating the parameter that has triggered the relevant law: for example, formation. If the explanation was to be purely test based this would be a retrograde step from the explicit reason but, as MALFIE provides a visualisation of the AIS signals, this short statement can cue the user to look at the parameter and value in the context of the overall picture and understand why it is an anomaly. Effectively, the implicit methods trade detail in the explanation for greater awareness of the context of the explanation. The third was the relative explanation, which provides a comparison of the probabilities of meeting a condition linked to a law.

Each of these styles of explanation can be achieved by using one or other of the ML techniques previously described as part of the second level of functionality. Explicit explanation can be achieved through the

---

9 B C van Fraassen, *The Scientific Image* (Oxford: Clarendon Press, 1980).

decision-tree algorithms. These represent Hempel's laws as branches and specific conditions as nodes, which lead to specific identities or priorities (leaf nodes). However, their very specificity can detract from the understanding of the explanation unless the user already understands the laws that link conditions and events.

Implicit explanation also depends on the decision-tree algorithms but rather than using the tree structure it uses a metric of value, such as information gain or the Gini coefficient, to identify the most significant parameter.

Relative explanation uses clustering or density distributions to measure the relative position of vessels' parameters. Each cluster or peak in a distribution would relate to a particular anomaly, pattern or priority. So, identifying which cluster or part of a distribution a vessel fits into would explain why a vessel should be a particular priority compared to the clusters or peaks of other vessels.

With all three of these levels of functionality, and the individual process stages within them, training was key. This would be unsupervised training, using historical AIS data to create many sets of normal and abnormal (anomalous) behaviours for different combinations of vessel type, season, month and areas, then different examples of patterns of interest, followed by different sets of patterns of anomalies from which the algorithms learned what the key parameters were and which were higher or lower priorities based on the relative frequency of each set of profiles.

Proof of Concept

Phase 1 of MALFIE was about proving the concept to potential stakeholders and users. Whereas a lab-based test or demonstration could get away with clunky algorithms with data scientists or computer programmers running code manually, we needed something that was sufficiently interactive that a representative military stakeholder could at least see how they might use it. Before we could do this, we had to agree on the use case for the concept.

The first use case was for 'alert and recording' where anomalies were highlighted and recorded for use in presentation to other agencies. This 'evidence generator' would be the least challenging for both the underlying algorithms and the interface, as there was sufficient time to run the algorithms and manually check the outputs, and the users could have a greater level of system and AI expertise. The next use case was for 'early warning' where current or potential anomalies would be highlighted along with sufficient context (including prioritisation and explanation) to support potential action via planning. This posed a medium challenge for the algorithm as although it would need to run fast, there were still some opportunities to check the outputs, and the prioritisation could be managed through setting thresholds. However, it would pose a high challenge to the interface due to the need to ensure rapid assimilation of the outputs by an operational user – in other words, someone who is not an expert in the system and its AI. The final use case was for high-risk monitoring of operations, in other words triggering immediate actions. This posed the highest challenge for the algorithm as there would be a high potential for, and consequences of, false positive and false negatives. However, it would pose a medium challenge for the interface because it could be optimised for a user who is expert at these types of decisions.

It was decided that the early-warning use case was the highest priority, as national and geopolitical events at the time meant that there would be opportunities to rapidly exploit MALFIE. This decision made the choice of the underlying anomaly detection algorithms slightly less critical (as the outputs could be complemented by other inputs), but did place more importance on the clarity and simplicity of the explanation and visualisation.

Two stakeholder events were held to demonstrate and get feedback on the concept. The first event included a lieutenant from a naval innovation organisation who had been the operator involved in the testing of the previous anomaly detection system, a retired lieutenant commander who had extensive experience of making the types of decisions MALFIE could support and writing the associated doctrine, and a technical expert from Dstl. An initial concept was presented that provided multiple screens. First there was a control panel that provided a high-level 'situational awareness' visualisation of all the vessels in an area and explanations of the level of anomaly for the most recent and highest priority vessel. The second provided tabular data of each vessel's latest anomaly levels either chronologically or by priority based on anomaly level. Third was a visual comparison of different vessels' anomaly levels, and fourth showed graphs of trends in anomaly levels. The anomaly values and resulting explanation for these screens and visualisations were based on AIS data from Scottish waters in 2016.

This first stakeholder event demonstrated that our three hypotheses were correct, albeit with the caveat that they were correct only when looking across a range of different types of anomalies and patterns. We used the feedback to design a second iteration followed by extensive unit, system and integration testing. This culminated in a second, much larger proof-of-concept demonstration, attended by the same individuals at the first event plus several more senior officers and civilian staff from various defence and security organisations. The event featured a presentation of the technology and concept, a demonstration of the application by the development team and 15 minutes of 'free play' by the stakeholders, which can sometimes be a nightmare but worked very well this time.

There were the usual comments regarding the things that were deliberately left out due to time and budget constraints, reasons why it might not be used by certain stakeholders, and the 'if only you could do XYZ' suggestions. These comments can be helpful, mainly as evidence to support what we already knew and wished to address in future work.

What was most useful was that the demonstration really landed with one of the stakeholders. Their organisation was facing new challenges caused by the changing geopolitical situation, and they had been given the resources to take things forward. Also, they had already invested in a range of different anomaly and pattern detection algorithms which would need an explanation and prioritisation overlay such as MALFIE to allow operators to avoid being overloaded by them. Finally, the organisation had access to a large amount of data sources covering AIS as well as radar and intelligence sources that the anomaly detection algorithms plus MALFIE could use. This led to a phase 2, again funded by DASA but with this stakeholder organisation now being the identified future user.

Adoption

Phase 2 of MALFIE was a much more user-experienced and integration focussed project than phase 1, for obvious reasons. It began with a very useful statement about the problem MALFIE would be solving for them:

> Analysis of vessel tracking is undertaken by the operator on a case-by-case basis, requiring a manual check of a vessel's track. This consumes a significant number of person hours, and also means that it is unlikely that unusual shipping movements will be noticed. The ability to automatically flag unusual behaviours would be a useful addition.

This statement led to three key criteria for the capability: the widest possible search area covering all areas that might contain unusual shipping movement of interest, a primary purpose of spotting unusual shipping movements to pre-empt and prevent undesirable actions, and the use of scarce resources in particular person hours.

The first stage was to link up additional anomaly and pattern detection algorithms, which the user had funded, with the MALFIE overlay algorithms. Next was to upgrade the user interface to be much more robust and usable. Then came the challenging work of integrating everything onto the user's existing systems in order to access their data sources and output results to their operators. Finally came the process of training the application and testing it with live data. We were fortunate that several incidents had occurred in geopolitical hotspots for which the MALFIE capability was ideal. We had recent data and operator experience against which to validate the benefit of MALFIE in two use cases. The first was contemporaneous (but counterfactual) early warning for headquarter staff monitoring a surface picture to task available assets and pre-empt an incident. The second was post-event analysis, using MALFIE to discover any additional information about the incident.

We found that a typical operator would receive over 4,000 location messages an hour from AIS alone, and would be faced with around 300 vessels at any given time. MALFIE was able to identify the 10 percent of vessels that were the highest priority for continued monitoring and investigation, which was a substantial reduction in the workload of an individual operator. We were also able to demonstrate that, across the real-life cases, MALFIE provided warning of anomalous behaviour of a vessel 20 to 30 minutes earlier than the traditional monitoring process. However, in the historical event that featured extensive AIS spoofing, the user interface – in particular the overall visualisation – provided clear indications that spoofing was occurring, but MALFIE was not able to adjust its learning to compensate.

Again, there were many comments regarding additional features that would be desirable and tuning that could be done for particular geographies, but such features were for the users to take forward. Our task was effectively complete. The concept had been researched and proven in phase 1 and phase 2, and a working system had been implemented on the system of a user organisation and its performance validated. Alas, at this point MALFIE took a different trajectory to DUCHESS. Although both concepts were widely applicable, because DUCHESS had no single, unique use case or user organisation it had to be fairly simple and generic, so that it could be used easily by any potential user. In the short term, that made DUCHESS challenging; but in the long term, once we had invested in the application, we had a much larger potential client list.

By contrast, the short-term benefit of an identified user organisation for MALFIE meant that the application was very focussed on their use case. Given that the MALFIE user was a niche part of the defence and security sector, the MALFIE application then became very exclusive. Theoretically the MALFIE application could be genericised but, as a small firm, we had to choose between MALFIE and several other AI technologies we had developed to invest our limited resources. DUCHESS was already well on the way to becoming a commercial offering, so, after much internal analysis we decided against investing in commercialising MALFIE because many firms were leaping onto the 'explainable AI' bandwagon. Whilst we were unique in our AI being explainable to a non–data scientist or military operator, this was not a strong differentiator in the commercial sphere, where data scientists were in greater supply (if only because many computer programmers label themselves as data scientists). So our work on MALFIE as an application ended with its adoption by this particular defence and security organisation. However, as one of the first companies whose 'AI explainability engine' went all the way from concept to use, we had a lead in terms of understanding and expertise on which to build further work.

Development

The MALFIE application was developed to be used for early warning in an headquarters-style setting, that is for a team in a secure location with many information feeds and experts available to review and report, rather than for an individual in an operational environment. Providing the anomaly and pattern detection outputs plus the MALFIE priorities and explanations via text and visuals on the screen was sufficient. The operators had the time to read whatever came through either immediately or later, and had large screens that allowed them to zoom in and out, select individual vessels, and chart comparisons over time and across vessels. Urgent activity did occur, but the team size and access to stable computing capability and communications made things easier than in a tactical environment such as an armoured vehicle with three people each with very specific roles (driver, gunner, commander), minimal screen and keyboard space, and the need to simultaneously monitor the area through head-up or helmet-mounted displays, sighting systems, telescopes and binoculars for threats that could appear and have to be dealt with in seconds.

From around 2020 we started getting involved in an increasing number of research projects looking at the application of AI in these tactical environments across the land, air and sea domains. As before, the refrain of 'how can we trust the AI if it doesn't explain' came up again and again, not just from the military but also from the technology companies. We would then present our MALFIE work, as the organisations within MOD which had funded and used MALFIE had not shared their work internally.

We soon found that presenting MALFIE as an example of explainable AI was not hitting the mark with stakeholders. At the time we blamed this on some combination of 'not invented here' and a desire to be seen as cutting edge and thus disregard previous work. With the passage of time we realised the real reason had more to do with the different context of the specific MALFIE application we had developed compared to the tactical context plus the need for stakeholders to be involved in the development of an AI application to fully understand the potential benefit.

The similarity between our previous MALFIE application and the tactical application of explainability was that in both cases the operator had multiple AI algorithms providing outputs, as per the multiple AI anomaly and pattern detection algorithms on top of which the MALFIE prioritisation and explanation sat. More important, however, was the difference. In the case of MALFIE all the AI systems it was trying to prioritise across and explain provided equivalent outputs; in other words they were working in parallel. They were all outputting indications of anomalous behaviour or patterns, just based on different criteria. In the tactical space, by contrast, each AI application typically addressed different parts of a single process. For instance, one might detect a vehicle, another would determine what the vehicle was and whether it was the enemy's, the next AI would try to determine what the vehicle was doing, and the next might recommend a course of action or even potentially react automatically. Although the various AI applications might work in parallel, the outputs made up a sequence, at least in conceptual terms.

In the context of sequential AI outputs and tactical imperatives of speed and ease of interaction between the human operator and AI, two different challenges arise when trying to automate the AI explanation. When should an explanation be provided, given that explaining things takes time and might be a distraction? If an explanation is required, what is the best way to manage the human-to-AI interaction so it minimises distraction but maximises the speed and correctness of the operator's decision?

We developed a concept called MAD HATAS (Multi-Agent Dialogue for Human Autonomy Teaming with Adaptive Systems) to explore these two points. The MADM we developed provided a tactical-level equivalent to MALFIE in that it sat on top of multiple AI algorithms and acted as the AI team leader. In other words, it was provided with the outputs from multiple AI algorithms and then decided which parts to provide to the operator, and how. It was designed to be similar to a crew member but one that interacted purely by voice (similar to being on the radio), with occasional injects and pointers on the screen or head-up display for crew in a vehicle, for instance.

Our approach to MADM was to consider how human teams interact. There are two ways people in a team learn when to explain. The first is just the natural process of trial and error with a new colleague, which leads to a sort of 'experiential configuration': Major Smith likes you to explain if ABC happens, Colonel McDuff only wants you to explain if XYZ happens. So one element of MADM was a set of thresholds that allowed different users to define when they would like a full explanation and the order they would like the elements of the explanation to be given (situation, key parameters, implication or recommendation). For example, if the AI was being used to monitor the sky and recommend actions to take against a swarm of enemy drones, a vehicle commander might set the thresholds in terms of the range at which to warn about the swarm, indicate what the swarm is doing and recommend an action (MADM says "Red swarm is within 300 metres, appears to be conducting surveillance, recommend you remain concealed to avoid detection").

We also allowed the operator to set different thresholds for different 'push and pull' interactions. The push consisted of those things the operator felt were urgent, in which case MADM would interrupt with the information and inject a circle on the screen or head-up display to draw the operator's attention towards the anomaly (the drones swarm in the example above, for instance). Other things were defined as important but not urgent, in which case MADM would indicate with a tone in a headset or a symbol on the screen or head-up display to say "I have some information – let me know when you're free to listen". For all the other things, MADM would simply load up into its database, ensuring out-of-date information was dropped, and wait for the operator to ask, similar to how Siri or Alexa do.

A second, and more interesting, mechanism that people use to avoid overloading a colleague with explanations is to filter out anything obvious and focus on what they assess their colleague does not already know about. For MADM to do this required it to either plug into the operator's brain or to have some means of predicting what the operator might have worked out for themselves. We chose the latter of these by applying a rapid-learning agent called Red Mirror. This used multiple ML models to attempt to predict an enemy platform's next steps purely by using the data from the specific engagement it was in. In other words, the agent was provided with no pretraining but was configured to accept specific parameters about the enemy platform movements. It would then build multiple machine-learned models of the pattern of enemy platform behaviour as the enemy approached, all the time learning which models behaved best and slowly building a 'mirror' of the enemy platform's logic.

Given Red Mirror only used 'in-engagement data', which usually consisted of no more than tens of minutes of observation, one might expect it to not be particularly good at prediction. However, in most tactical situations the range of options on a second-by-second basis is sufficiently small that we could, within a few minutes, get a relatively high prediction accuracy (often over 60 percent). That may not be high enough to base decisions such as whether to shoot something down, but as a metric for whether something is obvious, it was adequate. Basically, if we saw the enemy platform do something that Red Mirror predicted, then it must have been sufficiently obvious to not need further explanation. However, when we saw the enemy do something that Red Mirror did not predict, then explanations were needed because it was not obvious.

This work was conducted in 2023 and involved demonstrations with military officers playing through a set of scenarios and interacting with MADM in a simulated environment. There remains more work to develop this for use in a tactical space as the AI needs to be sufficiently robust and trusted to be suitable for making split-second life-or-death decisions (for the user, the enemy and other others in the vicinity). It nevertheless shows what could be done.

Reflections

MALFIE benefited from many of the things we learned from DUCHESS. First amongst these was building in time for software integration and testing. There is never enough, but at least one can plan as much time as possible up front and manage the features that one includes accordingly. It helps to have a team of experts in each aspect. Some team members focussed on the underlying algorithms and concept of use, others focussed on their areas of expertise. Had a big defence firm led on the project, perhaps they would have taken the same 'best athlete' approach, or the need to hit targets might have caused them to do as much as possible in house. In our case, we were resource constrained so were happy to bring in experts from outside.

The tension in defence between using technology and finding solutions really came out with this project. A low point was being told by a professor at one of the universities paid to develop one of the underlying anomaly and pattern detection algorithms that the project did not have technical merit because the individual techniques were old. That highlights a bias shared by many in defence research and development towards 'innovation as scientific invention'. Innovation can be the novel application of old ideas in new ways or new areas, but when it comes to funding the new is more attractive. By contrast the use of existing technology in new areas might have benefits in terms of cost and robustness, which are factors not often taken into account. Those marking proposals need to be aware of what has or has not been done in the past and for which application. Unfortunately, a marker may only have one or two hours to assess a proposal and that does not leave them time to check whether what is being proposed is genuinely novel in terms of application, whereas determining that a technique is novel can be easier.

The final reflection from MALFIE regards the role of AI. Whereas DUCHESS was AI that improved the contribution of humans, MALFIE is an example of AI that allows humans to use AI better. MALFIE does not replace the human operator looking for anomalies or specific patterns of behaviour, but it does ensure that the human is not a bottleneck when making use of the mass of alerts that could be created by AI anomaly and pattern detection systems. One could argue that needing AI to enable humans to make the best use of AI becomes a 'self-licking lollipop'. As organisations transition to use AI whilst keeping humans responsible and accountable for the decisions, tools like MALFIE can act as the 'adaptor' between a potential AI future and the human present.

6

Red's Shoes: AI That Learns How Your Enemy Learns

You must not fight too often with one enemy,
or you will teach him all your art of war.
— Napoleon Bonaparte

Just walk a mile in his moccasins,
Before you abuse, criticize and accuse.
If just for one hour, you could find a way,
To see through his eyes, instead of your own muse.
— from 'Judge Softly' by Mary Lathrap, 1895

Key Takeaways

This case is about artificial intelligence (AI) created to understand how humans learn, after the previous ones that described AI that draws out what humans have learned (DUCHESS) and AI that explains other AI to humans (MALFIE). Red's Shoes is an AI application that learns how the enemy (referred to as 'Red') learns from experience, noting Napoleon's caution that their learning comes from the experience one gives them. The name alludes to the saying that before someone judges someone else they should walk a mile in their shoes, as Mary Lathrap writes. In normal usage this is an encouragement to build empathy through looking at things from another person's perspective. The purpose of the Red's Shoes concept was to use AI to help 'take a walk in the enemy's shoes' (Red teaming) without the constraints and biases that a person's culture and experience may impose on their interpretation of the route the enemy has walked.[1]

As with DUCHESS and MALFIE there was an initial user pull so it was easy to address the first three questions that help with climbing up the utilisation staircase described in Chapter 1: *What existing (human) process is the AI addressing or is similar to? How is it different from other AI? What does this AI do better, or allow one to do, that was not possible before?* Our decision to focus on the questions of simplicity and ease of use meant that the question of time to mature and integrate was taken care of as well. The key takeaways from the Red's Shoes case relate to risks and understanding military use.

Red's Shoes is a good example of successfully addressing the question, from the utilisation staircase, *What perceived risks does the AI deal with?* Consumers and legislators are quite rightly concerned about the risk of the training data introducing bias within AI applications. However, there are a multitude of human biases that have long been of interest to behavioural scientists. The idea that AI might compensate for human bias is relatively new and Red's Shoes is one example of AI that can be better at avoiding cultural and experiential biases than people are, so long as there is sufficient internal and ongoing testing and tracking when using the AI.

The second takeaway is how it addressed questions of *What risks does it introduce* and *What will be done to address the new risks?* The perhaps surprising answer is that it (helpfully) blurred the line between quantifiable objectivity and qualitative subjectivity. Quantitative metrics are often considered objective 'facts' whereas an individual's narrative response is often denigrated as subjective opinion. If a customer or potential customer gave a verbal opinion of a product, the content of that opinion should be considered their fact – they liked it or did not like it for some reason. The collection of many opinions with a profile of these views from worst to best and points in between is a fact about the population as a whole. This has become more accepted as websites started asking people to give ratings, possibly because the number of stars absolved people from reading the tone

---

1   A red team is a team whose objective is to subject plans, programmes, ideas or assumptions to rigorous analysis and challenge. Red teaming is the work performed by the red team in identifying and assessing assumptions, options, vulnerabilities, limitations and risks. In AI it can also mean simulating attacks, but this is not the meaning here.

and tenor of the comments themselves. Large language models (LLMs) have made it much easier to enumerate and encode narrative feedback and Red's Shoes was able to make use of this.

The final takeaway from this case is that the answer to the question *Do we really understand how the military user would use the AI?* can change over time, to the benefit of eventual utilisation. Perhaps the most useful thing about AI for users in the military is not that it can provide a solution or prediction (as was initially assumed), but that it can provide insight and understanding (an example, perhaps, of Moravec's paradox[2]). As was found in the MALFIE case, humans often do not provide true explanations. Instead, they often fall back on post hoc justifications which are usually based on the idiosyncrasies of their individual experiences. This is because a real-life situation may be so unique, and the response required so quickly, that true explanation is not possible. For someone trying to explain their understanding of an enemy's likely actions based on their culture and experience (for the sake of buy-in and effective planning), the differences to their own culture and experience make this even harder.

Context

Two of the foundations of management consulting generally, and strategy consulting in particular, are the use of evidence and logical frameworks to make decisions. Neither are new, but by the 1930s, when management consulting took off, business had become sufficiently complicated to require something more than just simple calculations of cost efficiencies and production rates. The Boston Consulting Group and McKinsey became famous for (amongst other things) their four-box and nine-box matrices, more correctly known as the Boston matrix and direction policy matrix, respectively. These were highly rational frameworks to enable managers to

---

2 In 1988, Hans Moravec stated "it is comparatively easy to make computers exhibit adult-level performance on intelligence tests or playing checkers, and difficult or impossible to give them the skills of a one-year-old when it comes to perception and mobility" (*Mind Children*, Harvard University Press, 1988). Some years later (1994), Steven Pinker made a similar claim that "the main lesson of thirty-five years of AI research is that the hard problems are easy and the easy problems are hard" (*The Language Instinct: How the Mind Creates Language*, W. Morrow, 1994). The scientific basis is that what humans consider easy, such as the sensory and motor skills of a child, is the result of millions or billions of years of development, whereas what humans consider difficult, such as abstract thoughts, are difficult because we have only been doing it for a few thousand years at most. Hence, AI is able to master the new things that we are still trying to master but still has much time to go to learn to do those things which we have become able to do subconsciously through evolution.

make decisions about their competitive strategies. Both were variations of what became known as the analytic hierarchy process defined by Thomas Saaty in the late 1970s. This, in simple terms, involved identifying criteria for success, weighting these criteria, scoring the options against the criteria and then, by combining the scores and weightings, arriving at a simple measure of 'goodness' for each option, thus allowing management to choose the best one(s). This is now one of many methods that fall under the category of multi-criteria decision analysis, each with particular approaches to the weighting of criteria and scoring of options.

The great attraction of these methods is their 'rationality'. Stakeholders have to agree and weight the criteria explicitly. The options can then be scored using quantitative evidence (usually) to back the scores up. The final result is just maths. However, in practice rationality can often be circumvented. Different stakeholders, whether they be different directors or managers of different departments, can favour different options. So they could give a higher weighting to those criteria they know their favoured option would get a high score for. Conversely, they could give a lower weighting to those criteria they know their favoured options would get a low score for. Of course, in some situations, the corporate board requires a particular answer and is using the process to provide a logical reason to the shareholders. So they go around and around the process fiddling the criteria, weights and scores each time until the undesired answer drops out. When I was using these methods, which I often did when my consulting company was starting out, we would ask upfront if the directors genuinely wanted to follow the process to decide what to do or whether they just wanted to justify what they had already decided to do. If it was the latter, we used the 'fiddled' criteria, weights and scores to demonstrate 'what needs to be true' for their preferred options to come out on top. This would often highlight what assumptions they were making (sometimes implicitly) which were driving what they thought was the best option, and where those assumptions might be unrealistic.

At the very end of the 1980s an alternative approach gained popularity. This was the naturalistic decision making (NDM) movement, pioneered by Gary Klein amongst others.[3] It was driven by various observations about how people actually make decisions, and how this was quite different to the rational methods. This difference was particularly marked in situations with

---

3 Gary A Klein, Judith M Orasanu, Roberta Calderwood and Caroline Zsambok, *Decision Making in Action: Models and Methods* (Ablex Publishing, 1993).

incomplete information and the need for quick decision such as firefighting and the military. Research into NDM eventually led to the recognition-primed decision (RPD) model which, in simple terms, involves a decision maker looking at the features of the current situation, picking out the previous experience that most closely matches the current situation and applying the course of action that worked in that best-fit previous situation. If the decision maker does not have a sufficiently close experience, then they start assessing courses of action and implement the first one that will provide an acceptable result.

Unlike the rational methods, with RPD the decision maker does not list and assess all the potential options, usually because there is neither the time nor the information available to allow a purely rational decision-making approach and still be able to act in time. Instead, they either go with what worked 'last time' (or whichever time was 'close enough') or go with the first thing that comes to mind that works well enough, as it is quick and requires less information. Expertise in the form of past experiences (including training) is the crucial fuel for the mental simulation involved in RPD. The decision maker must mentally pattern-match the current situation against their portfolio of situations and, if no match is found, mentally simulate potential courses of action and outcomes until something acceptable is imagined. However, a decision maker's portfolio of experiences is likely to contain a mix of success and failures (at least in relative terms), which means there is an overlay of lessons that they have learned.

This explains why two individuals from the same culture and with the same education can prefer different courses of action when faced with the same situation. Their experiences have 'biased' them to favour certain decisions over others unless there is conclusive evidence to the contrary. For example, some officers habitually went left flanking and others right flanking when conducting attacks during exercises. Similarly, some officers habitually tried to stabilise or 'go firm' on a situation before planning things in detail, whereas others would try to go with the flow of events in order to create or maintain momentum. When asked to explain their reasons, they mostly highlighted the problems they had faced by not doing so in the past. In other words, they had learned from experience and were now applying that learning to avoid their previous poor outcome.

Inspiration

We started to try to model the learning cycle in around 2011 as part of our work in the finance sector. A few years before, behavioural finance had become a growth area, building on the scientific work of people such as Nobel Prize winners Daniel Kahneman and Amos Tversky. Interestingly, their work had many parallels with that of Gary Klein of NDM and RPD fame, with whom they sometimes collaborated, but Klein's contribution to this area has not received the attention it deserves. James Montier produced one of the first practical applications of the behavioural finance theory in his paper 'The Seven Sins of Asset Management'[4] which identified various biases among fund managers and the differences in the benchmarks that each of them react to when learning.

A key parallel between military decision makers and fund managers is that in both cases decision making is a zero-sum game. When one fund manager makes money on a trade, by buying an asset cheaply and selling it later for a better price, then another trader must be losing money. Similarly, when one military officer makes a decision that goes well, their enemy must have made a decision that went badly, and vice versa. Having made a decision and seen the outcome, both then could learn from it but when and what people actually learn is asymmetrical. When things go well, they tend to assume it is due to their skill and resolve to repeat it next time with an implicit assumption of consistency ('I did well last time so I'm sure I will do well next time'). When things go badly, however, they look for the reasons and try to change what they do, with an assumption of improvement or bad luck ('I did badly last time but I'm sure I will do better next time'). These are the 'hubris' and 'humility' cycles, respectively. However, in the hubris cycle, having done well, the enemy (in the military case) is in their humility cycle, having done poorly. So the enemy is improving whilst you are staying the same which leads the enemy eventually to do better and enter their hubris cycle, whilst you do badly and enter your humility cycle. These interacting cycles are shown in Figure 6.1.

---

4 James Montier, *Seven Sins of Fund Management: A Behavioural Critique* (Dresdner Kleinwort Wasserstein, November 2005).

Figure 6.1: Hubris and humility cycles

We soon started observing the characteristic 'seesawing' in battlefield performance, caused by the interlocking but asymmetrical learning cycles of two opposing commanders, in many historical conflicts. Napoleon summed things up rather well: if you fight the same enemy too often, they learn from you and get better. People often forget that although Napoleon was undoubtedly a military genius, he had a fantastic tool for his ambitions in the form of the French revolutionary armies. Revolutionary zeal made these armies large and with troops who could be trusted to forage without deserting (one reason for the speed of advance of French armies). The opening of officer ranks to anyone based on merit rather than aristocratic birth increased the pool of officers and their quality. Tactics were changed to make best use of the size and nature of these new armies. It took several drubbings but eventually the imperial armies of Austria, Prussia and Russia embraced similar ideas. They never quite matched the French armies on any of these aspects, but they came close enough for their numbers and patience to wear France (and Napoleon) down over time.

Within this observable high-level cycle of learning from, and copying of, what the enemy did (what Robert O'Connell calls 'the iron law of symmetry' in warfare[5]), the pattern-recognition part of Klein's RPD becomes evident. The key elements of the patterns that were considered by the commanders were what went well last time (as per hubris and habit) or what went poorly last time (as per humility and learning). The Prussians underwent the most thorough change, possibly because of the chasm between the scale of the twin defeat of Jena-Auerstedt in 1806 and the expectation based on Prussia's past military

---

5   Robert O'Connell, *Of Arms and Men: A History of War, Weapons and Aggression* (Oxford University Press, 1990).

prowess. The architects of the transformation, Gerhard von Scharnhorst, Carl von Clausewitz and August von Gneisenau, were all middle-ranking officers. They had early experiences in the Netherlands, the Rhine and the American Revolutionary War, respectively, between 1770 and 1800. All three were involved in the disaster of 1806 and spent the years leading up to the 1813 'war of liberation' contributing to the transformation of the Prussian military in different ways. Interestingly the changes in the Russian and Austrian armies were driven at higher levels. For the Russians it was generals such as Mikhail Kutuzov and Barclay de Tolly who drove many of the changes, whereas for Austria it was Archduke Charles. There is a correlation between the scale and shock of defeat and the level of learning. The Prussian defeat was such a shock and catastrophe that only relatively junior officers were capable of learning and were given the space to implement. The defeats suffered by Russia and Austria were significant, but less significant compared to the expectations, so the learning was more measured and top down.

Around the time we were trying to quantify learning effects, we were also working with Gary Klein, which caused us to look again at military history to see if learning cycles might be overlaid on successes and failure. That yielded the observations about the Napoleonic war described above, amongst others. Then, soon after, an opportunity appeared for us to test whether this might be exploited in the planning of modern warfare.

Inception

In many defence R&D projects, the requirement for innovation is a bit of a reflex, something people assume they must need. For example, when the UK is involved in live operations, the focus is on rapid impact and exploitation. Genuine innovation, which by nature is risky and hence has a higher likelihood of failure, often only occurs during periods of peace. The end of the 2010s was a period of genuine interest in innovation, as reflected by the proliferation of organisations and projects with the words 'innovation' or 'hub' in their title, or the letter 'i' at the beginning of their name (an homage to Apple no doubt). Towards the end of 2018, along with many other firms with a history of contributing to defence R&D, we were invited to an ideas generation workshop in support of an innovation-focussed project organised by the Dstl. It included niche technology firms like ours, large defence contractors and firms of ex-military officers who did a lot of work on wargames, exercises and requirements.

One area of interest was operational planning and, in particular, wargaming and red teaming. We were lucky that we had been teamed up with individuals from one company that specialised in human factors and another company consisting of ex-military planners, plus a senior Dstl analyst. Given the diversity of backgrounds the discussions were very wide ranging. One thing that came out was the difficulty of a military officer from a western military tradition being able to understand the mindset of an adversary from a different culture. Might AI be of use?

We concluded that to take things forward we needed to be able to address three things. The first thing was to determine at what level of military planning, wargaming and red teaming it might be possible to know who the enemy commander is and have access to their past experiences. The answer to this would provide the context in which the AI would have to work. In particular it would affect the amount and nature of the input data, the content and nature of the AI outputs, and how quickly it would need to generate a prediction.

The second thing was what the personality or character of the AI application should be. Both DUCHESS and MALFIE were sufficiently back office in nature and offered new capabilities (using AI to enable the learning of lessons in the case of DUCHESS and the prioritisation of vessels of interest at speed and at scale in the case of MALFIE) that users could be trained, within reason, to make the best use of the AI. Red's Shoes, by contrast, would fit into an existing real-time and front-line operational process conducted by military officers (red teaming) where the AI's capability was to help do something better rather than faster or at larger scale. This meant the AI had to be more aligned with how the military officers would readily understand, trust and act on the AI outputs in real time, because there was not the luxury of time to assess the AI's outputs for later use (as for DUCHESS and MALFIE).

The third thing was to determine to what extent the underlying AI algorithm could handle a mix of quantitative and qualitative inputs about the commander's experiences in a way that avoided bias. In addition, how would the algorithm's performance be assessed and how might it test and explain its outputs to ensure user trust?

We decided to form a team to submit a collaborative bid to develop a proof of concept for what subsequently became Red's Shoes. We led, as we had the underlying algorithm (from our finance sector work), and the other companies provided the user perspective and human factors input.

The combination of the novelty of the idea, the existing algorithm that could be adapted, our track record of delivering successful proofs of concept and the multi-disciplinary team (including people with military experience) made our proposal very attractive to the project stakeholders, and we received funding to prove the concept.

Technology

The underlying technology was a proprietary AI algorithm that had been specifically developed to learn and predict quantified learning effects. The fundamentals were that it took as inputs the various situations previously faced by an individual, the expectation of performance based on the situation and the individual's previous experiences of the situation, and the outcome achieved compared to the expectation. The algorithm would then learn the cycle that the individual went through, from their hubris-habit behaviour to their humility-learning behaviour, and the factors that drove the swing from one to the other, to predict how well the individual would do in their next experience.

The factors that drove the swing from the hubris to humility cycles fell into two categories. The first was the benchmark that each individual was sensitive to. Some people react most to relative benchmarks such as their performance compared to their colleagues, whilst other people react most to absolute benchmarks such as simple success or failure (irrespective of how well others did). The second factor was the time window for learning. Some people learn from what they have experienced only in the recent past, whereas others are still affected from experiences they had decades ago. What made the algorithm the basis for an AI application, as opposed to just a data science tool, was that it had the ability to learn, updating its model of an individual each time they had a new experience. So, having identified an individual's cycle, their driving benchmarks and driving time window, it would check each time a new data point (experience) was input and update these factors if needed.

One aspect applicable to military use was the need to combine quantitative inputs, such as the number of combatants on each side in a commander's experience, with categorical and qualitative inputs such as climate, terrain, mission, expectation and outcome. Dealing with categorical inputs was not in itself an issue for AI. Natural language process techniques, and the more modern LLM technology, meant that even dealing with qualitative inputs was possible. The issue was simply ensuring that information from different sources about a commander's different experiences could be structured and the different parameters scored or indexed in a consistent way. The goal was to convert descriptive and qualitative inputs into inputs to the algorithm, so it could produce outputs that were independent of any bias caused by the culture, training and experience of western military officers – or, if not initially independent of bias, that the algorithm would learn to correct for bias introduced by the sources.

The key feature of the output from our algorithm was that it only predicted the 'next' experience of the individual it was modelling. At the time several firms purported to have AI that could 'predict red' (the enemy) over the longer term. However, what red does next is a function of (or a reaction to) what blue – the planning team – has just done, which in turn is a reaction to what red did last, which is a function of what blue did previously, and so on. So, predicting the enemy's (red's) longer-term actions requires predicting blue's longer-term actions also. However, as with the learning cycles, red's reaction to blue's actions is a cyclical result of red's last experience, and their last experience equals blue's last action. This cyclicality means that, over a sequence of blue and red actions and reactions, where both blue and red end up can be very non-linear. Hence, any small error in the early predictions of either blue or red can end up causing a massive error down the line. So we took the decision to take the last blue action, which is red's last experience, as fixed and only predict red's next move. We then wait for blue to decide what to do, feed that into the algorithm as red's next experience, and predict red's action after that.

Another unusual feature of our algorithm, compared to most others, was that it did not require many data points to start predicting; three experiences were enough for it to be able to work, although the more the better. It did, however, require the data points to be rich, in the sense that each data point needed to cover multiple aspects of the past experience. It also required the data points to be adjacent, meaning that it could start from an individual's most recent experience and then provide the previous experiences until it reached a gap. After the gap there is little point adding further past experiences, which is a function of the non-linearity mentioned previously. If there is a gap in experience, there is a break in the algorithm's knowledge of where the commander is in their learning cycle, so it must start again until it has the minimum three experiences to start working again.

Proof of Concept

The approach we took to the proof of concept was to structure the work to address each of the challenges we had originally identified, namely most suitable level of military planning (effectively the use case for the concept), the personality of the AI application (effectively the user interface or experience), and the ability of the algorithm to manage with mixed data as indicated by how well it could assess its own performance and show its reasoning.

With the help of military subject matter experts, we soon identified that operational level planning was a good place to start. We conducted nine interviews with serving and former operational-level C2 practitioners. They highlighted that a key activity in creating operational-level courses of action (COAs) is the operational estimate (a process by which a commander decides on the COA that will achieve the mission).[6,7] Red teaming and wargaming are listed as key techniques for stages 4 and 5 of the operational estimate, namely 'validate COAs' and 'evaluate COAs', respectively. These then inform the commander's selection of a COA (stage 6). Relevant inputs to these stages of the estimate are provided by a red team and a J2 red cell.[8] However, the practitioners interviewed suggested that in practice COAs are frequently not wargamed until after selection, and the red teams themselves are ad hoc rather than specialist. In addition, a review of the MOD's 'Red Teaming Guide'[9] showed that of the 37 activities relating to the red team and red cell,[10] and the 30 analytical techniques used in red teaming,[11] none explicitly sought

---

6 UK Ministry of Defence, "Campaign Planning", Joint Doctrine Publication (JDP) 5-00", second edition, chapter 2, section 4, https://assets.publishing.service.gov.uk/government/uploads/system/uploads/attachment_data/file/434557/20150609-JDP_5_00_Ed_2_Ch_2_Archived.pdf.

7 JDP 5 has been superseded by the NATO (North Atlantic Treaty Organization) Campaign Operational Planning Directive, which is nested into Allied Joint Doctrine for Operational Level Planning. JDP 5 is used here as the potential user community is currently more familiar with it and it shares the same principles and detail with the Allied Joint Doctrine.

8 A J2 red cell is an entity led by the joint intelligence staff at headquarters that focusses on the activities of potential adversaries and threats.

9 UK Ministry of Defence, "Red Teaming Guide", Chapter 4 – 'Applying red teaming to defence problems', https://assets.publishing.service.gov.uk/government/uploads/system/uploads/attachment_data/file/142533/20130301_red_teaming_ed2.pdf.

10 Ibid page 4-4.

11 Ibid page 3-10.

---

to understand how red changes how they respond to blue's COA as a result of their knowledge and experience of previous blue COAs.[12]

The feedback we received was that the risk caused by this limitation in practice is, ironically, magnified by western militaries' technological advantage. The rapidity and intensity with which blue can generate 'traditional' effects mean that red get a very rapid education in which of their plans and assumptions work and which do not. As a result of this rapid education, they rapidly learn and rapidly change. It was also suggested that blue, by contrast, suffers from the 'asymmetry of assessment', which is that it is easier to prove that a COA did work than prove that it did not, as the perception of it not working could simply be due to a lack of information. Hence, blue may find it hard to abandon a COA that has been, or appeared to be, successful in the past (in favour of a new or previously unsuccessful COA) and remains firmly stuck in the habit-hubris cycle. Red, on the other hand, 'enjoys' the benefit of being regularly pushed into the humility-learning cycle and gets better at dealing with blue's previously successful COA habits.

'Joint Doctrine Publication 5 – Campaign Planning' highlights that the initial failure of the 1999 air campaign in Kosovo was due to 'failing to note how the adversary had adapted to previous experience', adding that 'it is often the loser who learns most'.[13] Frank Ledwidge[14] quotes Col (Ret'd) Richard Iron and Col (Ret'd) David Benest as having made similar observations on the impact of failing to take into account what the various shades of red (enemy), white (neutrals such as civilians) and green (allies) in Iraq and Afghanistan had learned from previous coalition operations when assessing how they will respond to the next one. US and Israeli authors have made similar observations about Israeli experience against Hezbollah.[15,16] We also heard that the nature of modern militaries, with their hierarchies and training regimes, become wedded to what they do well, which is the organisational equivalent of the hubris and habit cycle. By contrast,

---

12 The "Red Teaming Guide" was replaced by the "Red Teaming Handbook" in 2021, but this document provides less detail on activities and techniques. https://assets.publishing.service.gov.uk/media/61702155e90e07197867eb93/20210625-Red_Teaming_Handbook.pdf.

13 JDP 5 page 1-13 "Operational-level planning for the Kosovo air campaign: inadequate appreciation of an adversary's perspective".

14 Frank Ledwidge, *Losing Small Wars: British Military Failure in Iraq and Afghanistan* (Yale University Press, 2011).

15 Joshua Gleis, "Ten Lessons Learned by Hezbollah", *Huffington Post*, last modified 2 February 2015, https://www.huffingtonpost.com/joshua-gleis/ten-lessons-learned-by-he_b_6564850.html.

16 Moshe Arens, "Lessons Learned from Hezbollah", *Haaretz*, last modified 23 October 2017, https://www.haaretz.com/opinion/.premium-lessons-learned-from-hezbollah-1.5447889.

---

the relatively ill-equipped adversaries of the last 25 years have no particular organisational or hierarchical comfort zone and so are more willing to adopt what seems to work. It could be argued that the Russian[17,18,19] and Sri Lankan[20] campaigns against Chechen and Tamil rebels respectively only resulted in successful (if bloody) conclusions after they learned to give up on their previous assumptions of how they should fight and account for their enemies' learning, in order to adapt their operational plans and effects.

The conclusion of the practitioner interviews was that the greatest utility of Red's Shoes would be in assessing potential blue COAs in the face of a known red commander, or potential red commanders. This was thought to be most relevant at the start of a campaign or operation, as Red's Shoes would provide a mechanism to take the operational design (described in terms of lines of operation, centre of gravity and desired end state, and decisive conditions) and develop the operational COA by choosing supporting effects (SEs) that are least susceptible to the enemy commander's learning from experience. It would allow the user to iterate the COA by defining and evaluating operational risks. The user could identify the elements of the COA with the greatest susceptibility to red's previous learning, explain the contributors to this risk and test alternative COAs. The robustness of the plan could be assessed by testing it against different red commanders, so a robust plan would be one that has limited susceptibility to any of the learnings from experience.

With the level and use case in which Red's Shoes should be applied identified, it became easier to determine the type of interaction and personality it should have as an application. As an operational level application, it would have to mimic the conceptual level that staff officers in divisional-level headquarters think at. This meant that Red's Shoes had to be able to take as inputs the 'effects verbs' used NATO to describe what is to be done and achieved, and a standard list of 'objects' on which the effects are to be applied. For the pilot we used the effects verbs from the UK Staff Officers' Handbook. The objects are a much wider set of entities (conceptual or tangible) that

17 Unknown, "The Other Side of the Coin: The Russians in Chechnya", *Small Wars Journal*, accessed 25 November 2024, https://smallwarsjournal.com/jrnl/art/the-other-side-of-the-coin-the-russians-in-chechnya.
18 GlobalSecurity.org, "Second Chechen War". Accessed 25 November 2024, https://www.globalsecurity.org/military/world/war/chechnya2.htm.
19 Murad Batal al-Shishani, "The Chechen Mujahideen and the War in Iraq", *Journal of Slavic Military Studies* 18, no. 4 (2005): doi:10.1080/13518040500355015.
20 Taylor Dibbert, "How Sri Lanka Won the War", *The Diplomat*, 2 April 2015, https://thediplomat.com/2015/04/how-sri-lanka-won-the-war/.

effects could be imposed on. So we created a range of different classes such as the 'players' (red, red's opponents, white, green), components of fighting power (moral, physical, conceptual), joint functions such as (but not limited to) command and fires, operating domains, equipment capabilities, civil functions, and civil assets.

These were brought together to create a rich view of an individual red commander's previous experiences. Each data point consisted of the type of learning (whether cultural, historical, training or combat), the effects and objects that the blue commander (or the red commander's enemy) sought to apply against the red commander, the effects and objects that the red commander reacted with, the outcome from red's point of view based on a simple index, operational specifics such as blue and red missions, terrain, climate and dispositions, and, finally, a wide range of measures of the forces involved at the level of people, armour vehicles and artillery pieces.

We then created three interface views for the proof-of-concept application. The first was a series of input sheets, one for each red commander and their experiences (as described above). The second was the campaign plan where the user could define and iterate a plan step by step based on the red commander they wish to test the plan against. This view also presented the experience-by-experience testing that the algorithm does to assess whether it is modelling the red commander's learning cycle accurately. This step-wise testing was a very important feature of the output. It would be relatively easy to get a good correlation by testing the Red's Shoes application against a historic red commander with 10 operational experiences by simply inputting the 10 experiences and getting an algorithm to model them. However, the step-wise test is more realistic of potential future accuracy because now only the minimum number of rich data points (in this case three experiences) are input and the algorithm can predict the commander's fourth experience. This allows a comparison of the predicted reaction in the fourth experience against the actual reaction in the fourth experience. The fourth experience is input and so the fifth experience is predicted, which is input so that the sixth experience is predicted and then input, and so on. Now, any correlation is a better representation of what might be achieved in practice as each experience is input and the next one predicted in turn.

The final view lists the decisive conditions (DCs) and SEs from the campaign plan and breaks down the prediction of how the chosen red commander will react (based on their learning) against each part of the blue plan. Crucially, this also indicated which past experiences were most significant in driving the prediction, which provided a simple explanation. The algorithm predicts the learning cycle for the red commander's skill and will, inspired by the 'skill versus will matrix' used in management.[21] The will to resist is drawn from a report published by RAND investigating the national will to fight,[22] but applied to the operational level commander. This is driven by the algorithm's assessment of the combination of past sacrifice and lack of options.[23] The skill to resist is driven by the similarity of this experience to past experiences. Both skill and the will of the red commander can swing from hubris to humility and back again, so the algorithm predicts where the red commander is on both those spectrums and combines the results to provide an overall 'risk due to experience' for each of the DC and SE within the blue commander's plan.

Military officers have asked whether skill and will are the same as the capability and intent they often consider. These are not quite equivalent. The skill predicted by Red's Shoes relates specifically to the red commander's decision-making performance, whereas capability, in this context, refers to the overall force capability resulting from the mix of C2, doctrine, equipment, training, logistics, availability and other aspects of military effectiveness. Similarly, the will predicted by Red's Shoes relates to the determination of the red commander to succeed or fail, indicated by the willingness to take risk or the resources placed behind a particular reaction. The intent focusses more on the specific mission aim or goal.

For the proof of concept, Red's Shoes was tested on three of the commanders of Chechen forces involved in the first Chechen war (1994 to 1996), namely Shamil Basayev, Ruslan Gelayev and Aslan Maskhadov. They were chosen because they represented commanders with a different culture and military history compared to western military officers, their enemy (the Russians) also had a different culture and military history, and there was a good number of sources (both western and eastern) that provided the greatest chance of balancing potential biases. They each had about a dozen military experiences, initially in the Baltics or Abkhazia, and culminating in the recapture of Grozny in August 1996. This meant that we could conduct

---

21 The Peak Performance Center, "Skill Will Matrix". Accessed 25 November 2024, http://thepeakperformancecenter.com/business/coaching/skill-will-matrix/.

22 M McNerney et al, *National Will to Fight: Why Some States Keep Fighting and Others Don't* (RAND, RR2477, 2018).

23 "We must win or die – Pompey's men have other options", Julius Caesar prior to the battle of Pharsalus, which he won despite being heavily outnumbered by Pompey.

step-wise testing (feeding in one experience at a time and getting the algorithm to predict the next experience) on around nine of their experiences. This led to pleasingly high correlations between their actual reactions in each experience and the reactions predicted by Red's Shoes: 80 percent for Shamil Basayev, 64 percent for Ruslan Gelayev and 75 percent for Aslan Maskhadov.

We had no idea at this point whether a correlation of between 64 percent and 80 percent was good enough to be of use in military planning. However, it seemed high enough to conclude that the impact of learning cycles on a commander's skill and will could be predicted with good enough data on the experiences themselves. Quite rightly, these results resulted in some scepticism amongst stakeholders. The key suspicion was that the passage of time had led to much more detailed information about the commanders' experience than would be available for contemporary commanders. In addition, there was the concern that the Chechen commanders were in an existential and wholly kinetic operation in 2019, whereas the types of operation Red's Shoes might be used in would likely be hybrid/grey-zone in nature and far less existential for the red commander. Finally, although we took a robust approach to testing the AI, we did not properly test how easily a military analyst might collate, enter and output data from the application.

We were, therefore, given an extension, with the challenge of demonstrating that Red's Shoes could still perform well when analysing contemporary commanders with non-kinetic experiences, and whether an analyst without detailed knowledge of the AI could generate these results within a reasonable time. The first job was to decide which contemporary commanders to choose. We started with a list of six Russian commanders of general interest to the stakeholders, which was whittled down to two. These were General Alexander Zhuravlev, who had commanded Russian forces in Syria and was, at the time, the commander of the Western Military District, and General Andrey Serdyukov, who had extensive experience stretching from Kosovo to Crimea, Ukraine and Syria, and was at the time the commander of the Airborne Troops.

The second job was to find someone representative of a military analysis who could create the Red's Shoes inputs and generate the outputs. Ideally it would have been an intelligence analyst, but none were available. Instead, a recent graduate who had joined the company and had an interest in the military was provided with a detailed 'how to' guide to identify, collate and translate sources of the Zhuralev and Serdyukov's experiences, was given simple instructions on how to input to and output from the proof-of-concept Red's Shoes application, and used a timesheet to record the effort for each stage of the process.

The results in terms of prediction accuracy remained encouraging. There were relatively few open sources on these commanders, and even the fewer experiences (around half a dozen for each commander). The experiences were a mix of grey-zone or hybrid operations (Crimea and Syria) rather than conventional kinetic ones. Nonetheless, the correlation between the predicted reactions and the actual reactions for each experience was around 66 percent for each commander. In terms of the effort involved, it took the analyst 100 hours of effort (realistically three person-weeks) per commander, with no prior experience and from a standing start in terms of the information available. When this effort was broken down, 20 percent was spent simply finding sources, 50 percent spent reviewing, and 30 percent spent inputting and running the Red's Shoes application. For an intelligence analyst who could call on existing intelligence on the commander, this could be reduced to about one week of effort per enemy commander. We estimated that using modern LLMs to help structure the available data and automate its input would further reduce that time, perhaps to two days of effort per commander (something we have subsequently demonstrated).

Adoption

> *Great advantage is drawn from knowledge of your adversary,*
> *and when you know the measure of his intelligence and character,*
> *you can use it to play on his weakness.*
> — Frederick The Great

By the end of the proof of concept and its extension (early in 2020), we were convinced we had something of value. We were not entirely correct. We certainly had something that worked in the narrow sense of being able to make predictions, but we struggled to find anyone who had both the responsibility for, and interest in, using it in the way we had demonstrated it. Then Covid-19 happened, which threw everything into the air as many military officers were seconded to Covid planning activities, and the entire country got used to working from home.

For nearly two years our Red's Shoes work was just something interesting that we were able to present at innovation events and conferences.

Indeed, it won the 'best paper' award at the International Symposium in Military Operations Research in 2021. Then in February 2022, the Russian invasion of Ukraine began. Red's Shoes seemed like an ideal tool to help Ukrainian commanders get the better of their Russian counterparts, and we wanted to offer it to the Ukrainians for free. Unfortunately, every avenue we tried failed. The UK government launched various initiatives to funnel supplies to Ukraine but any bid had to be a minimum of £10 million. This was great for offering large amounts of artillery guns and shells, drones, communications equipment or electronic warfare equipment. AI applications for operational planning are just not as expensive – even if we wanted to charge for it. We tried to reach Ukrainian decision makers directly via UK organisations with links to Ukrainian institutes. Alas, each wanted a £10,000 introduction fee just to pass us onto the next person, who then wanted the same, with no guarantee of being put in front of anyone in the Ukrainian military who could decide to use it. Given that we were offering it for free, this was obviously not a commercially viable strategy.

Then we were put in front of a retired senior officer, from the military of a leading NATO member, who could offer feedback on whether our AI was as useful as we thought. His response surprised us. Having worked with his country's main force headquarters (divisional level) during previous NATO deployments, he said that Red's Shoes would have been very useful to him. He also commented that the new commander of this force was very open to new ideas, so he provided an immediate introduction. From then things moved very quickly. The general commanding the force headquarters passed us on to his assistant chief of staff for operations, and within a few weeks we attended a major planning exercise involving the entire divisional headquarters.

There were several reasons that a Red's Shoes–type capability was of interest to this particular headquarters. First, the move to an 'audience centric' approach to operational campaign planning had led to a need for the commander to be informed of behavioural changes within target audiences. One key audience in the operational environment is the red force commander, and the ability to offer insight on their behaviour would open opportunities to increase blue situational awareness (SA), inform blue COAs and provide potential routes to metrics for operational assessment. Second, the headquarters had routinely operated within the sub-threshold environment, which brought a need to diversify the types of SA provided to the commander to cover not just physical effects but also cognitive effects.

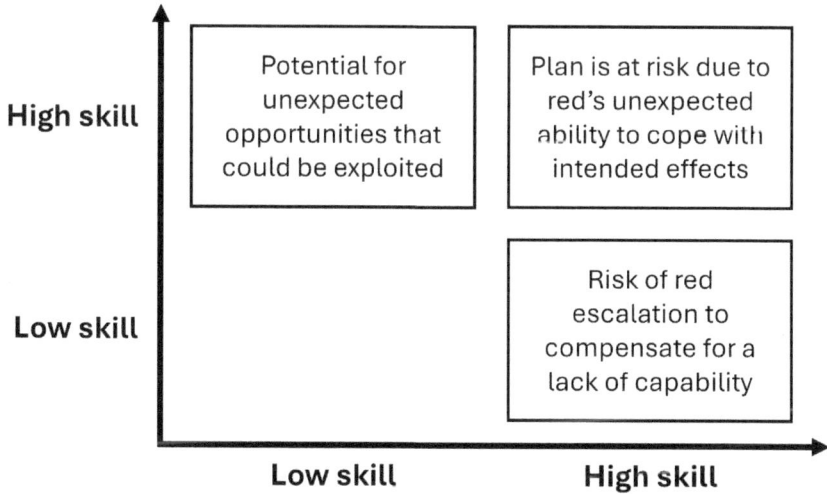

Figure 6.2: Predicted skill and will impact matrix for planned blue effects, and the resulting risk or opportunity for blue

Such cognitive effects include the skill and will of the red commanders and their potential escalation triggers.

For the exercise we demonstrated Red's Shoes by applying it to two contemporary commanders. We chose Admiral Alexandr Moiseyev, commander of the Russian Northern Fleet, and General Sergey Surovikin, who had been identified as an up-and-coming Russian commander based on his performance in the initial stages of the 2022 invasion of Ukraine. We were asked to focus on the modus operandi (MO) of the different Russian commanders and their potential escalation triggers (based on their skill and will) to inform both the choice of courses of action and how these courses of action should be portrayed through different means. A prediction of these commanders' next moves, in reaction to the blue COA, was not of explicit interest. Instead, the headquarters wanted to see if Red's Shoes could provide insights and understanding.

In working with the operations team, we realised that Red's Shoes' separate predictions of the red commander's skill and will in resisting the blue commander's planned effects provided more insight than originally imagined, as shown in Figure 6.2.

In the proof of concept, we had been using the Red's Shoes predictions to highlight DCs and SEs in the blue campaign plan, which fell into the top right-hand corner of the matrix, where the red commander had both high skill and high will that posed a high risk to the plan's success. These DCs and SEs could then be changed to those that fell into one of the boxes where the red commander had low skill, low will or both, which would increase the chances of the blue plan succeeding. We learned, first, that finding high-skill but low-will COAs was useful because these might go better than expected and could be exploited. We learned, second, that finding high-will but low-skill COAs was absolutely vital because in this case there is a great risk of unexpected escalation.

The other significant learning was how the headquarters made use of the internal step-wise testing that the application did on its predictions. The Red's Shoes analysis of General Surovikin's experiences (covering his Chechnya, Crimea, Syria and political/administrative experiences) resulted in a correlation between predicted responses and actual responses of 64 percent. This gave enough confidence to delve into the common features of success. Red's Shoes highlighted that these were collaborations with non-Russian military partners (such as Chechen militia, private military contractors, the Iranian and Syrian militaries), close coordination of ground and air assets (General Surovikin is, unusually, a former infantry commander elevated to air force general) and a focus on the key resource of the enemy prior to ground or air assault (in Syria this involved attacking the oil revenues of the Syrian rebels and ISIS). This analysis was conducted in early October 2022, one week before the Russian military launched the first wave of strikes against Ukrainian infrastructure using Iranian Shahed uncrewed aerial systems (drones). The media portrayed this as an attempt to terrorise or demoralise the civilian population in keeping with his description as 'General Armageddon'. However, the Red's Shoes analysis indicated that this was, in fact, General Surovikin's standard MO: collaboration with a non-Russian organisation (Iran) enabling a degradation of a key Ukrainian resource (NATO-supplied air-defence missiles, which must be used to defend critical infrastructure) prior to a coordinated air and ground assault currently prevented by Ukraine's intact air defence system.

By contrast the Red's Shoes analysis of Admiral Moiseyev yielded a lower predictive correlation (54 percent). Although Red's Shoes needs only a few data points to start working, these data points must be adjacent experiences, and the low correlation indicated missing experiences from the open source data gathered. It was possible to identify the key features of his preferred and successful MO (collaboration with foreign commercial organisations, demonstration of technical firsts, and a focus on extra-territorial action such as the Far North and the Black Sea) but the low correlation meant low confidence in Red's Shoes prediction of his skill and will in response to the exercise COAs.

During the exercise, Red's Shoes was mainly used as part of the daily briefing where potential COAs were presented to the commander, along with their pros and cons, and an initial recommendation from the planning cells on which to follow. The commander then gave their direction on whether to implement one or other of the COAs or to generate more. Red's Shoes was used to provide an additional set of pros and cons of whether the COAs provided greater opportunity than anticipated (high skill but low will) posed a higher than anticipated risk of escalation (low skill but high will) and what modifications to the COAs might lower the risks. Red's Shoes was highlighted as both unique and useful to the commander for this purpose.

However, it became clear that although the content of Red's Shoes was well received, its inclusion in the current process was a source of concern. Across a range of briefings to various individuals and teams at the exercise, the insights and outputs from Red's Shoes were met with great interest and we received requests for further detail. For those concerned with the smooth running of the headquarters process, though, Red's Shoes was considered a source of questions and options that would interrupt the flow of activities. On digging deeper, we concluded this was related to its presentation as an AI application as opposed to the view of an expert. Because the application of AI into higher-level headquarters was seen as new, it attracted many questions about the quality of the data it used, how it was trained and who had validated it. These are all appropriate verification and validation questions, but to ask and answer them for every COA within the drumbeat of headquarters processes was a challenge. Operational headquarters was much more used to incorporating different human experts (such as cultural, political, economic and historical analysis advisors) presumably because they were much more comfortable delegating the responsibility for qualifying these experts to the recruitment and contracting processes. Plus, many of these advisors were either former or serving military or civil servants so knew how to present and explain their views in ways senior military officers could easily assimilate and accept.

We considered the exercise as a success from the point of view of having Red's Shoes used by real decision makers in a realistic situation. But we clearly had some work to do in finding the most natural place for it to be used.

Development

Over the following 12 months we again presented Red's Shoes to various individuals and organisations, this time with the benefit of the insights from the divisional-level exercise in 2022. One such organisation was a defence think tank embarking on a US-funded study to explore the idea that the actions and strategies western countries employ now may influence the development of future adversary force structures. If this was true, employing the correct strategies now could influence a future adversary to adopt force structures that would be less formidable threats in the future, depending on the western country's own future force structure. If one assumed that organisations have similar learning cycles to individual commanders, then it might be possible to repurpose Red's Shoes to help understand how to influence a potential adversary's learning to end up in this ideal situation. The think tank contracted us to use Red's Shoes to explore three questions in relation to Russian force structure development: what are the relevant experiences from an organisational point of view, what are the patterns of their experiences and efforts to change, and, finally, what might have happened had the experiences they had learned from been different?

Five sets of inputs helped us understand the experiences the adversary had, and what drove their learning. The first input was the individuals at the top of the organisation. Whereas Red's Shoes had been applied against individual commanders, here we were looking at the defence ministers to help understand what made an experience. Defence ministers did not necessarily take each relevant decision, but their leadership influenced the direction or trend of decisions and strategies that tended to occur at the start and end of an experience at this strategic level. The second input, and perhaps most significant at that level, was the campaigns fought. We started with the first Chechen war, as it was the first campaign following the dissolution of the Soviet Union in 1991, and added the second Chechen war, the Georgian war, the Syrian intervention, the 2014 Donbass conflict, the North Caucasus insurgency and the first year of the invasion of Ukraine in 2022.

The third set was the various reform initiatives and actions undertaken. These were the combined effect of the experiences from operations and the activities of the defence ministers at the time. These three inputs highlighted a pattern (identified by the Red's Shoes) that the windows of experience at the strategic level tended to consist of two stages: a first stage of low-intensity or counterinsurgency combat capped with a second stage of a short period of conventional warfare. These two stages were not necessarily in the same geography, but they seemed to represent a slow or long period of experience building and force reformation during the first stage, culminating with a test of the changes during the second. The experiences from each of these time windows then led to the next reformation or reorganisation of forces.

The fourth input was the force structures announced immediately after the experience window. These represented the target force structures as reported, not necessarily the force structures achieved or at full strength. This input also took no account of the quality of equipment or training. It simply included the intent following the experiences and the learning from the most recent window of operational experience.

The final input was a set of two parameters generated in a workshop with military subject matter experts: the expectation of the outcome based on media reports, public statements and level of resources allocated to the operation, and the perception of the actual outcome, based on subsequent activities that implicitly demonstrated the view of the outcome. This was equivalent to Red's Shoes assessment of the outcome of individual commanders' experience but applied at the operational and strategic level.

Once the application was fed and run, it highlighted several learning drivers and responses. The intensity and outcomes versus expectations of each campaign experience very clearly influenced the balance between different types of regiments. This influence was most obvious in the swings between the investment in motor rifle regiments versus surface-to-surface missile regiments, and swings between investments in tank regiments versus light infantry regiments (such as Russia's airborne forces).

We then conducted a 'what if' exercise. The Russian force structures at that time were a reaction to the initial experiences in Ukraine. In testing, Red's Shoes was able to predict accurately how those initial experiences had led to the current structures. We then ran several alternative scenarios regarding the experience gained by the Russian ground forces in 2022. These scenarios varied the nature of the campaign (low intensity versus high intensity) and the outcome (more successful versus less successful). These indicated that the current force structure could have been significantly different had the nature and outcome of the first year been different.

These predictions are not shared in this book, as they are sensitive. They do, however, reveal some dilemmas for decision makers seeking to use these types of AI applications. First, is there such a thing as a preferred force structure that an enemy can be influenced to adopt? A country's force structure very often drives the type of conflict it fights; for example, the Iraqis fought differently in the first Gulf war from how they fought in the second. Over four decades in defence, in the UK there have been many individuals who have served in the Falklands, the Gulf, Kosovo, Bosnia, Northern Ireland, Iraq and Afghanistan – covering the whole gamut from peacekeeping through low-intensity conflicts to high-intensity ones. From their stories, it is unclear which of these types of conflict could be considered 'preferable' either as conflicts to fight or ones to deter. It is also unclear what criteria politicians would be brave enough to choose in order to make that call.

Second, how ethical would it be, for instance, to equip and train a country in order to ensure its enemy learns particular lessons rather than succeeds? In an ideal world the help provided would achieve both a victory for the country as an ally and the 'correct' learning for the enemy. In the real world it is highly likely that helping an ally to win would lead to the adversary learning the 'wrong' lessons from the point of view of creating the force structure the country would prefer to fight.

Nine months after this study, things came full circle. The highest-level headquarters of the organisation that had used Red's Shoes in its planning exercise set up a new red team. The name did not do justice to its role, which was to do more than act as red in exercises but also to provide a greater level of challenge to assumptions throughout the organisation. We briefed the members of this new team in a workshop organised by the think tank we had recently worked with on the various things we had used the Red's Shoes concept for. They proposed a hybrid problem: how would the Russian military modify its doctrine at the operational level based on its experiences in Ukraine? Like the think tank study of potential force structures, this was about something higher level than the learning cycle of individual commanders. However, the doctrine they might adopt generally comes from things tried by individual commanders in the previous conflict, which means there is some accumulation from, and aggregation of, the specific learning cycles of individual commanders.

Again, the results are sensitive and require regular updating as new experiences are added and the horizon for future doctrine moves ahead. The process, however, shows how AI applications can help bridge the gap between business-as-usual analysis and operational planning. We began by testing whether our understanding about learning cycles, which the underlying algorithm uses, apply to doctrine. We did this by focussing on how the experiences of Russian tank armies in World War Two led to the immediate post-war doctrine for tank armies, and also the post-Stalin modification of the doctrine (on the basis that Stalin's view of certain generals may have suppressed certain aspects of learning until after his departure). The Russian tank armies tended to have a single commander for most of the war, although they operated under different fronts and hence front commanders for different operations. Nevertheless, Red's Shoes was able to account for these different constraints (in the same way it adjusts for the constraints of force ratios, missions and terrain, for instance) and correctly identify the key drivers of the subsequent post-war and post-Stalin tank army doctrine. Following this, we applied the Red's Shoes insights on drivers to the tank army experiences from the Ukraine war to predict the next iteration of Russian tank army doctrine, and the variations that might occur due to the particular skills and preferences (or wills) of individual tank army commanders in the future.

The outputs of this work could have short- and long-term impacts. In the short term, it could be used when developing scenarios for planning and exercises, and to drive the behaviour of the red commander in exercises to ensure that the command teams are given the most robust challenge when learning how to fight future enemies. In the longer term, these scenarios and red behaviours could form the basis of the threat against which future defence technologies may be procured, as the west moves from the post–Cold War capability-based procurement process to the threat-based procurement process.

Reflections

Red's Shoes began within a couple of years of both DUCHESS and MALFIE but took longer to gain traction. This came as a surprise because the original concept was created to address a challenge posed by Dstl, and the team included ex-military experts from exactly the type of organisations that could use it. However, the type of person who leaves the military to work in a defence research business as a subject matter expert thinks differently about adopting AI applications from a serving military officer. From the point of view of an AI practitioner, an ex-military officer looks and sounds little different from a serving military officer. However, an ex-military officer no longer suffers the time constraints, reputational risks and processes that constrain a serving military officer from pulling through apparently useful applications.

Red's Shoes only got traction when we found organisations (as opposed to individuals) that had already bought into the underlying problem we were addressing (such as the think tank) or had a mandate to deliver something that required this problem to be overcome (the newly established red team). Without the Russian invasion of Ukraine, it would likely have taken both these organisations much longer to reach this point – which means that there was little that could have done to make progress any earlier. We just had to keep Red's Shoes in the front of as many people's minds as we could until the time was right for someone to pick it up. That is not the type of strategy that any normal AI firm would follow and was only possible in our case because we had other technologies that were making money in the meantime.

The other surprise was the extent to which clients did not want to use the application but just wanted to receive the insights. This was different from DUCHESS and MALFIE, which were very much intended to be something people would use as part of their processes. We suspect this is because both DUCHESS and MALFIE used AI to do things that humans could, conceptually at least, do themselves. The AI just allowed people to do things quicker or more easily. Red's Shoes was specifically designed to overcome human limitations in understanding and predicting red commanders due to cultural and experiential biases. So it did not naturally fit into what people could conceive of doing themselves, initially. Whether they were in the headquarters exercise, the think tank study or the red team project, people preferred Red's Shoes to be an external asset that generated insights for them to use, effectively an asset-based consulting model. Over time, consulting frameworks do get adopted by clients. The famous Boston Matrix and its McKinsey competitor the Directional Policy Matrix have become staples used by internal strategy teams and MBA curricula. Eventually Red's Shoes may also be used in house, but for the moment it is our consulting tool.

Finally, Red's Shoes was an excellent example of how AI-generated insight can be useful, as opposed to the type of AI-generated answer that people usually think of. People often think that 'AI in defence' is about an autonomous vehicle's manoeuvres controlled by AI, or sensors where the AI identifies whether something is a target. Admittedly, we originally thought of Red's Shoes as driving planning decisions in an equivalent way ('choose this course of action not that'). In the divisional-level exercise, and then in both the think tank and red team studies, we found the most useful outputs were the unusual insights on commander skill and will, the force structure trade-offs driven by campaign experience, and how individual commander experiences combined to form doctrine.

7

DR SO:
Machine-Speed Experience

Key Takeaways

DR SO is an artificial intelligence (AI) application that uses deep reinforcement learning (DRL) to develop tactics for attack and defence. It is, probably, the closest to what people often imagine when they hear the phrase 'AI in defence'. Whilst DUCHESS, MALFIE and Red's Shoes were all developed specifically to leverage or assist human operators and processes (lessons learned, prioritisation and explanation, and red teaming, respectively), DR SO's outputs can be used as part of autonomous systems as well as a tool to inform human decisions. It is also different in the type of human decisions it seeks to improve. As Daniel Kahneman writes, there are two types of thinking: 'system 1' is fast, instinctive and emotional (very often driving reflexive actions), and 'system 2' is slow, deliberative and logical. The DUCHESS, MALFIE and Red's Shoes applications all contribute to system 2 thinking, although Red's Shoes does so by using AI to determine the experiential drivers and patterns of system 1 thinking in a target commander. DR SO, by contrast, uses AI to mimic the experiential drivers of system 1 thinking so that the AI can learn and implement the appropriate reflexive actions.

The core of the application is that the user defines the physical capabilities of platforms such as ships, aircraft or drones, and DR SO learns tactics to allow each side to maximise rewards that relate to an objective. The military go through this process via training, exercises and ultimately fighting wars, where initial tactics are tested and improved. However, this process can take years and cost lives. The slow pace at which technology and tactics were developed in World War One to overcome trench warfare is an example of this. Applications like DR SO allow this process to occur much more quickly.

As it fitted within the relatively popular idea of AI that enables autonomy, there were accepted answers to many of the questions in the utilisation staircase from Chapter 1: *What existing (human) process does the AI address or is similar to? What does the AI do better? What is the level of ease of use? What perceived risks does it deal with or introduce? What will be done to address those risks? Is there enough time for it to mature and be integrated into existing systems?*

However, DR SO was a 'typical' use case for AI, which meant that answering the question *What is the difference to other AI?* became a significant driver of utilisation. The key takeaway from this case is for users to beware of terminology that confuses AI techniques with AI applications. In the finance world, many executives have a knack for rapidly picking up and using terminology in ways that make them sound knowledgeable but which can give the entirely wrong impression. Unfortunately, many individuals within the consulting industry rely on the same skill with similar unfortunate consequences. Defence is no different. As AI is clearly 'hot' at the moment, many people are jumping onto the bandwagon with minimal deep knowledge or experience and, as a result, misleading military officers about what has been done (and can therefore be acquired quite quickly and inexpensively), versus what could be done (but at some cost), and the pros and cons of these alternatives.

The second takeaway is that much value can be gained from focussing on the question *Is the AI as complex as needed, or as simple as possible?* and applying the AI in a simple way rather than the most sophisticated and technically integrated way. Clearly, if money and time were no object why would anyone not get the gold-plated solution? Defence tends to be resource constrained and needs to operate at pace, hence it typically accepts the 80 percent solution in time rather than the 100 percent solution too late. The phrase 'let's crack on' appeared to be a motto for the British military for a time and is rooted in this trade-off. There is no reason why AI development and implementation should not follow the same approach, but the military can sometimes be seduced by complexity if there is a promise of capability. This may be due to a combination of the desire to gain an advantage over the enemy and the relative ease of judging advantage in terms of theoretical performance rather than implemented performance taking into account reliability, robustness, ease of use and other factors that depend on long-term testing and implementation.

The final takeaway is that when addressing the question of *Do we really understand how the military user would use it?* one should consider whether the 'use' could be at several removes. DR SO is a good example of how the contribution of AI could take place at different parts of the industry-defence enterprise. Again, there is often the assumption that AI should be used in the platform and connected to a network. That definitely applies in some cases but there are plenty of cases where the AI can be used to generate non-AI modules for use within platforms in operations. This is similar to deciding at which command level to give certain delegations and freedoms, which, of course, relates to natural intelligence. However, although in the west there is a tendency to maximise freedom of thought and action (within certain constraints) for human decision makers, there is no reason that should automatically be the case for AI. Clearly, trust and ethics are a major part of this calculation.

Context

Drills have been a feature of military training for millennia. The Spartans and other Greek city states drilled their citizen soldiers to march and manoeuvre in unison. The Romans did the same, adding constant practice with weighted wooden swords and shields so that the real things felt easier. The foundation of modern military drill was the practice pioneered by Maurice of Nassau to train his men to conduct the 42 separate actions needed to load and fire the heavy firearms of the late 15th century. One of the main benefits of this drill was that it turned complex actions into something close to a reflex so that the soldiers could continue to function despite the incoming fire from the enemy and without being distracted by the death and destruction around them. As warfare became more mobile and fluid, and smaller units of soldiers were expected to fire and manoeuvre across the ground, battle drills such as the section attack drill and ambush drill became common. Veterans of the infantry battles in the Falklands conflict attest to the 'almost electric effect'[1] that seeing their section drills working had on the troops.

Discussions of tactics have been somewhat looked down upon. This is illustrated by the phrase 'amateurs talk tactics, professionals talk logistics'.

---

1 Ken Lukowiak, *A Soldier's Song* (Secker & Warburg, 1993).

This quote is normally attributed to General Omar Bradley, although there is no attributable source and it may well have been an anti-Patton jibe. Sources aside, this statement is true when professionals have learned and practised their tactics so well that the tactics work against most enemies, so they do not need to talk about tactics anymore. Logistics, on the other hand, are always difficult and variable in war. This is due to the enemy unhelpfully blowing up roads, railways, bridges, arms depots and factories. This is true in conventional warfare as much as guerrilla warfare. Distance and climate can pose challenges too. So British professionals in recent wars (the Falklands, the Gulf, Iraq and Afghanistan) have always ended up having to talk about logistics and rarely had to worry about tactics.

Tactics become a worthy topic of professional conversation when the nature and technology of warfare change in such a way as to make the tactics of the time obsolete. The best-known example is World War One, when machine guns, artillery, trench systems and barbed wire meant the typical tactics for conducting attacks led to the mass slaughter of infantry and the redundancy of cavalry charges. What is often forgotten, however, is that the US Civil War was the first to show the dangers of advancing across open ground, due to the accuracy and range of modern rifles. It was also one of the first wars in which extensive trenchworks were used in battle (as opposed to just sieges). The Franco-Prussian War just five years later taught many of the same lessons, although von Bredow's 'Death Ride' at the Battle of Mars-La-Tour provided an example of a 'successful' cavalry charge that hindered the learning of lessons about the vulnerability of cavalry in modern war.[2] The Boer War showed that these lessons were relevant even when a major power (Britain) was facing a far less powerful foe (the Boers).

Interestingly, World War One illustrated the two routes to overcoming tactical obsolescence. The Allies resorted to technology in the form of the tank, which they started to develop in early 1915 and first used in the battle of the Somme in September 1916. However, it took a year for them to achieve success with tanks at a local level, at Cambrai in November 1917, and another nine months for tanks to have a decisive impact, at Amiens in August 1918. Ultimately, the success of the tank was a mixture of having the technology, deploying it in sufficient quantity and refining the tactics. The Germans developed tanks also, but primarily focussed on innovative tactics in the form of their Stoßtruppen (literally 'shock troops' but translated to 'storm troops'

---

2 Bryan Perret, *At All Costs: Stories of Impossible Victories* (Cassell Military Classics, 1998).

in English). They began innovating in early 1915, around the same time as Churchill formed the Landship Committee that kicked off the development of tanks. The first successful employment of Stoßtruppen came in October 1915 in the Vosges Mountains, with subsequent small-scale successes in the early stages of Verdun in June 1916.[3] The large-scale employment came in March 1918, after Russia withdrew from the war, which allowed units to be taken out of the line for retraining. Germany's Operation Michael broke the stalemate of trench warfare for the first time in four years, and allowed them to launch three more offensives. Ultimately the Allies managed to contain the German advances and, with the benefit of the fresh US divisions that had arrived, were able to launch the counteroffensive that began at Amiens (along with the massive deployment of tanks).

Fast forward a hundred years, and we have seen again how the mismatch of tactics and technology can lead to heavy casualties. The large losses experienced by the Russians during the invasion of Ukraine since 2022 is, arguably, the best recent example.[4] However, this war also illustrates that, under the pressure of combat, soldiers will learn which tactics work. One has to allow, of course, for different attitudes to what the best metrics are for what works. For some countries it is casualties that count most, whilst for others it may be missions accomplished. Discussions with those who with first-hand experience of this war suggest that each side took around five weeks to develop tactical and technical counters to new systems introduced by the other side. However, during the weeks or years that it takes to develop better combinations of tactics and technology (as with tanks and storm-trooper tactics in World War One), many lives may be lost. Using AI technology to investigate and develop better tactics at speed could be of significant benefit.

Inspiration

Since around 2015, we have become involved in several maritime air-defence projects. The work has included investigating a range of historical missile attacks against ships so that we could generate realistic scenarios in which to test potential future concepts. Eventually, these projects have expanded to cover surface warfare and one of the cases we looked at was the attack on the

---

3 Bruce Gudmundsson, *Stormtroop Tactics: Innovation in the German Army, 1914–1918* (Praeger Paperback, 1995).
4 Jack Watling and Nick Reynolds, *Meatgrinder: Russian Tactics in the Second Year of Its Invasion of Ukraine* (Royal United Services Institute, May 2023).

Royal Saudi Navy frigate *Al Madinah* in January 2017.[5] Initially it was thought it had been rammed by a suicide speedboat launched by Houthi rebels in Yemen, but it was subsequently determined that it was remotely controlled. This ship was the first of its class, built in France in 1985 and upgraded by France in 2013, and at the time of the attack was operating in open water as part of a military operation (the Saudi intervention in the Yemen against the Houthis). Yet a single explosive-laden speedboat was able to approach and explode, killing two sailors and causing significant damage. The Houthis were supplied and trained by the Iranian Revolutionary Guard Corps Navy (IRGCN), who made extensive use of swarming tactics involving both speedboats firing rockets and missiles. So it was seen as a simple evolutionary step to upgrade from boats with remote-controlled explosives to remote-controlled missile-firing boats and combine this with swarming tactics. It would be another simple step to replace the remote control with AI control of the boats.

This has led to us becoming involved in some projects looking at how to deal with AI-controlled swarms. Red Mirror, discussed in the chapter on MALFIE, was one such project, and sought to rapidly build a 'mimic' of the enemy's AI purely from observations of its reaction to manoeuvres. Concepts such as these require an AI-controlled threat platform against which they can be tested. We were looking for AI that controlled a drone to conduct swarming attacks, meaning multiple drones that would surround a target and conduct a suicide or missile attack simultaneously from multiple directions. Unfortunately, at that time (2019) no one had, or admitted to having, this kind of multi-axis, simultaneous-attack swarming AI. Many companies claimed they had AI for swarming drones, but what they typically meant was many drones moving in close formations, which is really flocking. So we took the decision to create one for ourselves.

Inception

All the applications described so far in this book were the result of some need identified by the UK Ministry of Defence (MOD) via one or other of its agencies. They presented us with a problem or an area in which they had an interest in investigating and we came up with ideas to solve the problem

---

5 Sam Lagrone, *Navy: Saudi Frigate Attacked by Unmanned Bomb Boat, Likely Iranian* (USNI News, last updated 20 February 2017). https://news.usni.org/2017/02/20/navy-saudi-frigate-attacked-unmanned-bomb-boat-likely-iranian.

or explore the issue with a new or existing technique. If our work showed promise, it became the basis of a technology or system that the military used.

DR SO was our first experience of finding a problem, and investing our own money from the start, to develop a solution that we would use in our defence work ourselves.[6] This is standard for a tech startup, except that they need to raise funds from investors. We were not a startup in the strict sense, as DIEM has been around since 2011 and was spun out from a consulting company established in 2002. As a result, we were in the luxurious position of having our own money and not needing to look for outside investors. This meant, on the one hand, that we did not have the challenge and stresses of looking to give investors a quick return. On the other hand, it meant we did not have the experience of investors to guide us. We consequently fell into our own 'hubris cycle' of the type our Red's Shoes AI might have identified.

The problem we identified was the result of our observation of how quick it can be to create capable AI – at least compared to most defence equipment. Between 2015 and 2022, we did a lot of research into the potential for AI to improve maritime air-defence. Some of the applications we had worked on were tested during naval at-sea exercises, and the feedback validated our work. We were even interviewed for a *Janes* article.[7] The problem was that if we could do it, so could an adversary. A further problem was that an adversary might have fewer ethical constraints and hence might find and deploy a solution even quicker than we would.

This got us thinking: could we use AI to analyse and predict an enemy's AI? Indeed, could we use it to assess whether our AI was being predictable? We put forward funding proposals to research these questions which, eventually, led to the Red Mirror and Multi-Agent Dialogue module projects mentioned in the chapter on MALFIE. The problem we had to address first, however, was that no one had an AI 'agent' against which we could test our outputs. Or perhaps no one wanted to admit to having something whose predictability they would let us test. As no one could (or would) share their 'red agent' AIs, we created an application that could rapidly generate red AI agents for each of our different projects.

Western navies have long prepared to defend themselves against an adversary, such as the IRGCN, using speedboats, fast attack craft (FACs) and

---

6   In the case of DUCHESS we invested our own money to develop the application after initial MOD funding. In the case of Red's Shoes the technique was originally developed internally as part of work in the finance sector.
7   Richard Scott, "Machine Speed Warfare: UK Tests Naval AI Decision Aids in ASD/FS-21 Exercise", *Janes International Defence Review* (July 2021).

fast-inshore attack craft (FIACs) to swarm capital ships. Then the Houthis used a single, Iranian-supplied, remote-controlled speedboat to successfully strike a modern Saudi frigate. It did not require any leaps of imagination to assume that AI-controlled speedboat and FIAC swarms would soon be a threat. Having already built a reputation for AI in maritime warfare, we decided to begin our Red AI agent work by looking at AI for swarming. This initial focus led to the name 'DR SO', which stands for 'Deep Reinforcement learning Swarming Optimisation'.

Technology

When work on DR SO began, in 2019, the concept of swarming was enjoying a resurgence. The war in Ukraine may be the first major war where drone swarms have been used frequently and at scale. As a concept, however, military 'swarming' has thousands of years of history as highlighted in a number of RAND papers from the early 2000s, which was the last time swarming was a 'thing'.[8],[9],[10] Unfortunately swarming was seen as a selling point for technology (not unlike AI now) and there was a lot of marketing spin. This confused people as to what swarming actually is and, hence, what a swarming algorithm does.

As is sometimes the case, the flexibility of the English language does not help. The verb 'to swarm' means "to move somewhere in large numbers", according to the Oxford English Dictionary, but it can also mean "to move in or form a swarm". The US Merriam-Webster Dictionary defines it as "to move or assemble in a crowd" and also "to beset or surround in a swarm". These differences are small but important. One definition focusses on surrounding a target as the key feature of a swarm, and another has movement and numbers as the key characteristics. This contradiction has allowed many companies to label the algorithms that kept their drones flying in a group, or a formation, as swarming algorithms. However, we would label these as flocking, herding or formation-flying algorithms.

In a military context, the RAND definition of swarming is a good one: "autonomous or semi-autonomous units engaging in convergent assault

---

8 Sean Edwards, *Swarming on the Battlefield: Past, Present and Future* (RAND, March 2000).

9 John Arquilla and David Ronfeldt, *Swarming and the Future of Conflict* (RAND, September 2000).

10 Sean Edwards, *Swarming and the Future of Warfare* (RAND, May 2005).

on a common target".[11] This was the one we used. We started by looking at various swarm intelligence (SI) techniques. These are bio-inspired algorithms, many of which recreate the behaviour of flocks of birds and schools of fish. Particle swarm optimisation (PSO) is an SI technique that looks for the best solution and has been used in route generation. Swarming involves multiple agents each finding a route to a specific position around a target, so this seemed like a viable option. PSOs use a conceptual swarm of particles, where each particle tests an initially random solution. The particles then communicate how successful their solutions are and they converge on the optimum. Unfortunately, it only worked if the target was passive or followed a predictable path. If the target reacted to the swarming agents, and had comparable manoeuvrability to them, the PSO created behaviours that looked like children chasing a ball around a field.

We needed a technique that had some capability to predict and pre-empt the target's reaction. We also wanted it to make better use of the fact that the swarm outnumbered the target. That led us to a range of multi-agent DRL algorithms. As with PSOs, DRLs test random initial solutions and then converge on the optimum. The difference is that PSO test many solutions at once whereas DRLs learn over many random attempts, improving each time. As we trained our multi-agent DRL we initially saw the same 'children chasing a ball' behaviour that we saw with the PSO, but then we started seeing improvements.

The key to making DRLs work is having a good reward scheme, also referred to as a reward function or reward structure. This is a method of scoring the end result of the DRL's random attempts. The DRL learns by trying random actions to see which leads to the highest rewards. It then builds on the highest-scoring actions to improve the total rewards further. In other words, a DRL is a sophisticated version of trial and error conducted at machine speed rather than the speed of experience. In the case of the swarming behaviour we wanted DR SO to learn, the reward scheme was based on how close the red agents got to the blue agent, how evenly spaced they were around the blue agent and how completely the blue agent was surrounded. For the blue agent, the reward scheme was driven by how long it could avoid being surrounded by red.

The initial training scenario featured four red agents trying to surround a single blue agent, in the presence of two obstacles. All the agents, whether

---

11 John Arquilla and David Ronfeldt, *Swarming & the Future of Conflict* (RAND, September 2000).

blue or red, had the same speed and manoeuvrability. With a moderate level of training, we noticed that three of the red agents would chase the blue agent and the fourth red agent would hang back to try to catch blue if it got past. This seemed odd until we realised that the blue agent had learned to head towards the obstacles in order to break up the red formation, creating gaps it could slip through.

When we allowed the algorithm to keep training, red eventually learned to create a formation to block the blue agent from getting close to the obstacles. Red then used this formation as a barrier to push the blue agent into a corner. As both sides had the same amount of training, red's four-to-one advantage meant that blue could not get away.

Proof of Concept

The first use of the DR SO swarming algorithm was as part of the Red Mirror project. Western navies had already put a lot of effort into developing doctrine to counter the type of FAC and FIAC swarms that the IRGCN had. It seemed a natural progression for potential adversaries to replace the human crew of these FACs and FIACs with AI that could swarm their targets. Red Mirror used rapid machine-learning techniques (also called 'low-shot learning') to watch the red drones as they reacted to the blue ship's initial manoeuvres, and develop a version of the red swarm's AI logic. This would then allow the blue ship to predict what the red swarm would do next, and hence avoid being surrounded. DR SO basically acted as the red swarming FACs' and FIACs' controller.

The main purpose of the work was to prove Red Mirror's ability to predict how the red swarming drone (controlled by DR SO) would change direction each second. However, we also wanted to demonstrate that what DR SO was learning was superior to what could easily be developed. So we sought out a range of counter-swarming tactics, which we turned into behavioural heuristics (rules of thumb) for the blue agent to use. Most of these came from counter-FAC/FIAC tactics developed by western navies. However, we also included a heuristic based on the behaviour of African antelope, taken from a nature documentary and turned into a simple heuristic. This involved the defending blue agent waiting till the red swarm was close, clustered and moving fast, and then jinking out of the way. We then compared how long the blue agent avoided being surrounded using these different heuristics versus how long they survived using DR SO's DRL-learned behaviours in a computer simulation of an attack.

When we compared the performance statistics of the various heuristics, we found that the two that kept the blue agent from being surrounded the longest were the tactics that DR SO had learned and the antelope's 'jink' manoeuvre. When we ran the visualisations we saw that DR SO had, in fact, learned the same jink manoeuvre that the antelopes used. The difference was that DR SO had learned this tactic over the course of several hundred thousand machine experiences over two nights, whereas the African antelope had learned over millennia of evolution. They proved superior to the standard naval drills for one important reason. For the jink manoeuvre learned by both evolution and the DRL to work, the implementation has to be very precise. Such precision is possible with AI control and, in the case of an antelope, superior agility. That level of precision is much harder to achieve with a ship under human control, let alone a ship under the control of a stressed and fatigued human, which is likely to be the case in combat.

Development

DR SO was originally an algorithm to generate red AI agents for our own use within research projects aimed at investigating or predicting red AI threats. After having seen how well it could learn advanced tactics for both offence – for example, swarming – and defence – for example, counter-swarming – we realised we could develop it as an application in its own right. It could then be more easily applied to use cases that required more realistic conditions in which to train the DRL and learn the tactics.

The first such use case was a development of the counter-drone scenario in the context of providing guidance to a human operator. Here is a scenario: the sensors in an armoured vehicle in a hidden position looking out over the front line suddenly indicate a swarm of drones approaching. How do the crew react? If the vehicle has been seen, they need to either move away or shoot down the drones to avoid an attack. However, what if the vehicle has not been seen? Moving or firing would give away its position. If it fires before being noticed, it would need to bring down all the drones very quickly to limit the chance of them attacking or relaying their position. If the crew relays its position to an air defence unit, it might be helpful but if it unmasks too early, that signature might be picked up by the enemy and invite counter fire.

The key to making the right decision is to determine the swarm's intent. Are they doing a surveillance sweep or are they targeting the vehicle? This is where DR SO comes in but in a way that generates insight to a decision maker rather than controlling an autonomous vehicle. By training DR SO as if it were the red swarm, it could learn different red swarm tactics. It learned red surveillance behaviours and red swarming attack behaviours. Then, when the red swarm is detected by the sensors on the vehicle, DR SO is run by taking in the current situation (the blue and red positions and directions) and outputting what it would do in these different cases. By comparing what the red swarm was seen to be doing to what DR SO would do if it were the red swarm, it was possible to assess intent. If the red swarm's behaviour closely matched the attacking behaviours learned by DR SO, then they were assessed as having seen the blue agent. If the red swarm's behaviour closely matched the surveillance behaviours learned by DR SO, then they were assessed as having not seen blue yet, but looking for it. If they matched neither set of DR SO behaviours, then they were assessed as just being in transit.

This way of using DR SO was tested with a group of military officers, as part of a project focussed on human-autonomy teaming (HAT). It was most striking how quickly they incorporated the outputs of the DR SO comparison into their decision making. Within a day they had settled on the extent to which they would just monitor the DR SO outputs whilst they did other tasks, and the point at which they would take over from the AI and focus on the red swarm.

The second use case was also related to HAT but, rather than acting as a decision aide for the human, provided the behaviours used by the autonomous vehicles acting in concert with a crewed platform. This combination of human-crewed platforms and autonomous platforms working together was called an autonomous-collaborative partnership. One key challenge in such a collaboration, when the autonomous vehicle is AI controlled, is to ensure the humans on the crewed platform know what the autonomous vehicle will do in a particular situation, such as when the vehicle detects incoming fire before the humans do. The default assumption was that the vehicle had to be able to communicate to the crewed platform what it was going to do, so that the crew could accept or veto it (part of the mechanism of ensuring meaningful human control). However, such communication and acceptance could slow things down or overload the crew.

As mentioned in the comparison of DR SO behaviours to naval tactics to counter-swarms, the way in which the military are trained to react to a threat must take into account the ability to implement in a stressful situation. Hence, the military have many simple drills, standard operating procedures (SOPs), and tactics, techniques and procedures (TTPs), which they are trained to implement almost as a reflex. The section assault drill, for instance, is one of those that every young soldier learns in basic training. The ambush drill is another. Pilots have SOPs for when they hear the missile warnings in their headphones, and naval crew have drills for when a missile launch is detected. The key to all these is that everyone knows what to do and, crucially, everyone knows what everyone else is going to do. It seemed that the behaviours learned by DR SO could be converted into autonomous-agent SOPs (AASOPs) or autonomous-agent TTPs (AATTPs), which would allow the humans to know what the AI-controlled autonomous vehicles are about to do without slowing things down.

The project involved several technology companies, each with different algorithms that represented competing solutions. Each company was invited to pick one of three vignettes to demonstrate their algorithm's functionality. The first vignette involved the launch, forming up and formation flying of four platforms (one human-crewed platform with three autonomous drones). The main test of the algorithm was to see how well it maintained formation whilst avoiding obstacles and terrain. The second vignette involved creating a plan to avoid a potential threat using the different platforms' sensors. Here, the algorithm's speed and quality were tested. The third vignette involved a rapid reaction to an unknown threat firing on the formation. The main test was to see the extent to which the algorithm minimised the number of hits on the platforms in the formation, given that it would not know whether the threat was a man-portable aid-defence systems, anti-aircraft artillery or surface-to-air missiles systems until after it had been fired on. We chose to focus on the third vignette as it was the best demonstration of DR SO's outputs, and its outputs could be also used as the solution for the second vignette.

Training the DR SO DRL highlighted several interesting points. We took a curriculum-training approach, where we train it in simple scenarios first, and then provide it with increasingly complex scenarios. For the simplest scenario where a single drone had to learn to evade a single threat, DR SO learned to conduct a figure-of-eight manoeuvre. Military stakeholders we presented this to suggested that although violent 90 degree turns were standard threat evasion manoeuvres, doing a figure of eight was risky for two reasons. First, it is challenging for a human pilot to consistently make the most of what ever manoeuvrability a crewed platform has. This is less of a constraint for the autonomous platform, however, as the drone is more manoeuvrable, and the AI can consistently exploit this. The second reason was that a figure of eight keeps the vehicle in the same area and that leaves it vulnerable to attack from a second threat.

So we then increased the complexity by having multiple threats from different positions. This generated the more realistic behaviour of regular turns whilst moving away from the threat. We called this behaviour 'jiggy' internally, and the description rather stuck. We were told subsequently that that the military term is jinking, but we had already applied that to the antelope behaviour in the maritime counter-drone use case.

For the final training exercise, we combined multiple threats with a full formation. The DRL learned four different behaviours depending on where the threat was located relative to the formation's heading. The first we called 'sacrifice', when DR SO sent the drone that was closest to the threat towards the threat itself. At first this seemed like an error but then we realised that the DRL had learned that if the threat was very close it was better for one of the drones to soak up all the threat missiles or shots and let the others escape. The second we called 'flee', when the DRL got all the drones simply to move away as quickly as possible. It learned to do this in situations when the threat was far enough behind the formation for the crewed platform and drones to get beyond the maximum range of the threat before it hit them. The third we called 'mask', when the drones conducted jiggy manoeuvres in a direction that put them between the threat and the crewed platform. The DRL did this when the threat was ahead and close enough to hit the crewed platform. Effectively, it learned to use the drones to draw fire away from the crewed platform whilst exploiting the greater manoeuvrability of the drone to avoid being hit themselves. Finally, DR SO did the basic jiggy manoeuvre when the threat was ahead of the formation but when the crewed platform was able to evade on its own.

We were able to use a clustering approach to isolate the situations in which the DRL implemented these different behaviours, and these became four different AATTPs or AASOPs. Each was defined by the range and bearing of the threat compared to the formation's heading, which made the collaboration really easy. The only thing the human crew had to know was what and where the threat was. From that, they knew exactly which of these four manoeuvres the drones would implement.

This approach has the additional benefit that the advanced DRL itself does not need to be on the autonomous vehicle itself, thus reducing computer power and the risk of the AI falling into enemy hands. After every mission the DRL can be trained with whatever threat situation has been faced that day. The AI can then be run offline to improve its behaviours and the updated A ATTPs (rather than the AI itself) can be uploaded to the autonomous vehicle before the next mission. This is procedurally very similar to how communication encryption keys are handled.

Adoption

DR SO began as an internal tool we used in projects researching ways of countering an adversary's AI and autonomous systems. Eventually it became the key asset in projects looking at how human and AI teams can work together to exploit the AI's ability to learn from experience. This progression led to considering how to best leverage DR SO to deliver operational impact.

The first step was to develop it into a much more usable application, with a graphical interface, user guide and training manual. This was not easy but the process was conceptually straightforward. We focussed on the technical requirements and contracted a large, global IT firm with a good track record of delivery of defence software to implement it so that everything was done to a high standard.

The second step was to create licence agreements. This was far more complicated as there were so many options. If DR SO was to be used by individuals in another organisation, how do we ensure they use it correctly? This is both an ethical and a processual issue, and quite challenging to cover in a licence agreement. Should we follow the software-as-a-service model, which was, at the time, favoured by tech firms as it made scaling up so much easier? How would we price it? We decided to be prepared for any combination, as we concluded that different defence clients would favour different arrangements. We had a law firm that specialised in intellectual property draw up several types of agreements.

The third step was to explore the market and find clients. The previous use cases had highlighted two potential types of client. The first was the various defence warfare centres responsible for the creation of doctrine (including TTPs and SOPs). We had little trouble finding people in such organisations interested in seeing what DR SO could do. We struggled, however, to get anyone to commit, as it involved a change to the current system of doctrine development. Prior to 2022, there was little incentive to rapidly develop new doctrine. After Russia invaded Ukraine, the incentive became clear, but it was easy enough to exploit the lessons the Ukrainians were learning. Once the war ends, perhaps there will be greater pull for changing doctrine in order to use AI applications like DR SO.

The second type of client was defence manufacturers. These faced the challenge of developing systems that could be used by militaries with different types of doctrine, against adversaries with different doctrines. The timing was good as the proliferation of drones meant there was a new set of small and highly manoeuvrable threats to protect against. In addition, projects for fifth- and sixth-generation fighters were being launched, so there was funding for future-focussed technologies. The challenge, however, was that whilst defence firms had much experience in contracting for services and subcomponents, the nature of DR SO made it fall between the cracks. It represented a new type of capability, but the firms did not have experienced users who could write the requirement to allow their commercial team to procure it. Contracting us to run DR SO for them meant they would not grow their own capability but would depend on us. Moreover, both the data needed and the output generated would be both highly classified and very commercially sensitive.

For these reasons it took several years for a leading defence firm to contract for the use of DR SO. It involved multiple proposals, each of which was turned down because it failed to address one or another of their concerns. Finally, it was realised that it was not possible to have everything at the start and that any limitations in the initial project could be addressed through the use of follow-on contracts to put everything they desired into place. The plan was to start with a one-year contract for a proof of concept, leading to a white paper demonstrating how the use of an AI application to learn adversary doctrine could improve the performance of defence systems. This would result in DR SO being implemented on their internal systems but run by us to ensure it was used to its maximum capability. This approach had the additional benefit that it would give users the chance to mature their understanding of how the outputs of the AI would fit into the concept of employment of the military capability overall. In order for AI developers to take more responsibility and accountability for AI ethics, it is necessary for the technical team to understand the military context in detail. Follow-on contracts would then involve rolling out the application for use by the internal development team, with us just providing support and consulting as needed.

Reflections

In some ways DR SO follows the pattern of many technology innovations of the past. We experienced a problem and implemented a solution for our own internal use, and then found that others could benefit from it and built a commercial offering. The messaging technology that underpins e-mails and the internet followed a similar path, having initially been developed internally by DARPA (as DARPAnet) before becoming a global solution. However, for us, it took six years to go from the initial idea in 2015 to getting the first external user, which was twice as long as for DUCHESS and MALFIE. This is partly because using AI to develop blue and red doctrine is a very niche capability, and partly because we are not a tech startup focussed on a single product.

In addition, some particular issues slowed us down. We underestimated the importance of coming up with a good tagline that both describes what DR SO does and differentiates it from what applications do. As it was originally just an internal tool, it was fine to say it creates swarming algorithms. The moment we tried to interest people outside the company, we ran into the problem that 'swarming algorithm' is often used to describe 'flocking'. We tried to explain why our swarming was different to their swarming. In hindsight we should have described it as a 'counter-swarming' application and changed the name from DR SO, but we were too attached to its roots and original conception.

DR SO was the first time we developed something using one of the more advanced AI techniques, namely DRL. Prior to that, we focussed on leveraging simpler AI techniques, usually with an ensemble approach (combining techniques), to generate insights from sparse and narrative data. The swarming / counter-swarming case neatly aligned with the DRL's ability to generate its own training data. However, converting the DRLs outputs to simple behaviours (the AATTPs) made it much more useful. This highlighted the benefit of the military and technical community actively seeking opportunities to simplify rather than relying on computing power to handle the complexity internally.

Where the complexity of DR SO did prove useful was in its use further back in the defence supply chain than our other applications. Whereas DUCHESS, MALFIE and Red's Shoes all produced outputs for direct use by military operators, DR SO's first external users were defence equipment developers. They used DR SO to create swarming and counter-swarming algorithms that could be integrated into the software used in defence equipment. This is the defence equivalent of using ChatGPT to generate art, books, songs and other 'copy'. This use of AI in defence often gets forgotten amongst the concerns about 'killer robots' that draws the most attention. A more useful concern is how the people using AI in this creative way can be (or should be) held responsible for the resulting performance of the defence equipment. Prior to the arrival of AI, defence manufacturers largely relied on military operators to implement the laws of war and rules of engagement to ensure ethical outputs. We found that ensuring the AI technical team understood the detail of the military's concept of employment for the AI-driven systems, was key to them understanding the technical requirements they would need to meet to fulfil their responsibility for AI ethics.

Part III: Lessons

8

Testing

Why Me?

One of the common themes in these cases where artificial intelligence (AI) is successfully brought into defence use is that many of the key people involved are formerly with the Ministry of Defence (MOD) or military. That would mean that they already understand the type of evidence needed to make a good business case and could more easily work with the military users to fit into their operational process. This is not a guarantee, of course, but it helps massively. In many of the cases where AI fails to have an impact, it has been because there was insufficient understanding between the AI developers and the military users. Sometimes this has led to the AI developers creating something that is just not usable in the military context. Other times it has led to military officers being seduced by the claims of the developers (or their salespeople) and then being disappointed by the results. As this book is intended to inform military officers who set the requirements, procure and implement AI, a section on testing AI will help bridge the gap of understanding and avoid disappointments.

The foundation of AI use in defence is the need to confirm the fundamental level of trustworthiness or dependability in the AI's outputs through testing. In the early days of AI, when presented with its potential, military officers would raise strategic questions about ethics and whether they can trust it. The policy framework has matured since people started developing defence AI with the UK now having guidance on both AI ethics[1] and dependability[2] (a synonym for trustworthiness) plus approaches to

---

1 UK Ministry of Defence, *Ambitious, safe, responsible: our approach to the delivery of AI-enabled capability in Defence – Annex A: Ethical Principles for AI in Defence* (UK MOD, 15 June 2022).
2 UK Ministry of Defence, *JSP 936 Dependable Artificial Intelligence (AI) in Defence*, February 2024.

assurance.[3] Figure 8.1 illustrates how the UK ethics and dependability policy documents relate. The UK's principles of AI ethics are shown in grey to the left whilst the aspects of the Joint Services Publication (JSP) on dependable AI are in black, on the right.

The principle of reliability is broken down into three properties which must be demonstrated (it must be reliable, robust and secure). The JSP states that dependability relates to how much a user can rely on three functions (safe, secure and correct operation). The JSP also links the demonstration that the AI is reliable and secure (on the left) to the functions of correct and secure operation (in the middle), respectively. Finally, as illustrated on the right of Figure 8.1, the JSP states that the amount the user relies on the AI should be commensurate with the confidence they have developed in the system. The point is that this confidence must be supported by adequate evidence. Therefore, the three key things from these two policy documents tie together: evidence, correct operation and reliability. One needs evidence of correct operation to demonstrate that the AI is reliable enough to be used ethically. Testing is the activity that generates the evidence, and the military should be involved as early as practical to ensure the activities and evidence are relevant and appropriate for the context.

Practicalities

Many different terms are used when discussing the generation of evidence of AI performance (be it on its own or as part of a system) and hence whether the AI is trustworthy and dependable. Such terms include metrics and measures of merit (both objective and subjective), baselining, benchmarking, accuracy, correctness, unit testing, system testing, integration testing, acceptance, assurance, and verification and validation (often combined as V&V). Many academic and professional publications discuss in detail how to measure the performance of AI. The following summary provides the key ideas needed to form a proper view on the trustworthiness of any algorithms or applications to help anyone involved in developing, procuring or using AI.

Broadly, there are three aspects to consider. Initially, the most appropriate type of measure to use needs to be identified. This is largely

---

3 UK Ministry of Defence "Assurance of Artificial Intelligence and Autonomous Systems: a Dstl biscuit book", last modified 1 December 2021, https://www.gov.uk/government/publications/assurance-of-ai-and-autonomous-systems-a-dstl-biscuit-book/assurance-of-artificial-intelligence-and-autonomous-systems-a-dstl-biscuit-book.

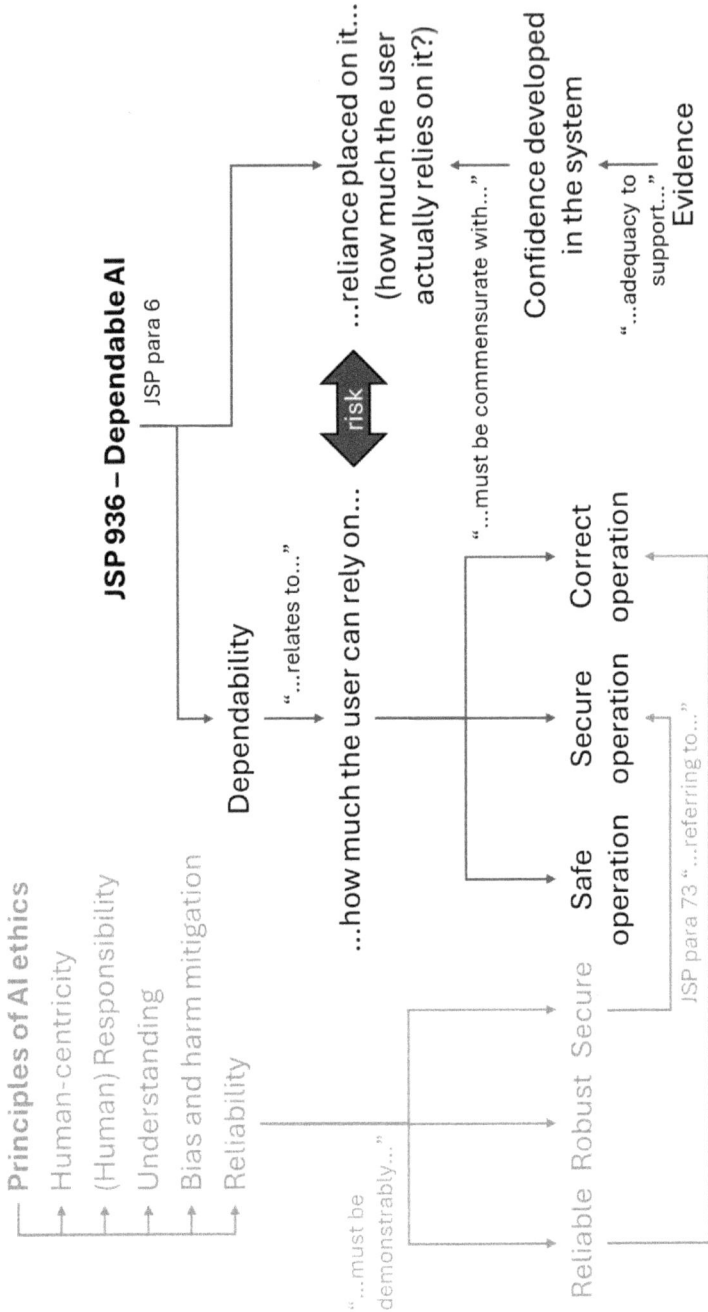

Figure 8.1: Relationship between the UK principles of AI ethics and JSP 936 on dependable AI

driven by the type of problem the AI is addressing. Typically, AI techniques are separated into those that deal with classification problems and those that deal with regression problems. Put simply, classification AI seeks to apply a class label, or put something into a specific category. Regression AI seeks to produce a value that sits on a continuous scale, such as speed, direction or temperature.

Second, and sometimes ignored, the baseline or benchmark against which to compare the AI's performance needs to be determined. This is not necessarily an AI-versus-human test; it could be AI versus luck or AI versus a simple heuristic. This aspect should also consider issues such as the split between training data and testing data (and how this distinction varies between, say a machine learning (ML) approach versus a reinforcement-learning approach), and the extent to which the data are unbalanced.

The third aspect is to identify how this AI improves the socio-technical system that it is intended to sit within, whatever its performance versus the baseline or benchmark. Although the idea of a socio-technical system has been overshadowed by terms such as human-machine teaming (HMT) or human-autonomy teaming (HAT), it captures the broad operational process, which may feature several sets of HATs that would ultimately be improved through the use of AI. The key here is to build a hierarchy of metrics that start with the basic measures of AI performance and go up through the HAT measures (including aspects of dependability and trust) to demonstrate the measure of operational performance that allow the potential benefit provided by the AI (which then allow overall ethical risk to be assessed).

Classification versus Regression

In AI and ML terms, classification means identifying something as belonging to one or another category and labelling it as such. In many cases this gets reduced to a binary problem, that is the AI decides whether something is or is not in a single category. An example is AI designed to detect armoured vehicles (is this a tank – yes or no?). Creating a cascade of binary AI outputs can lead to apparently sophisticated outputs, such as 'if yes, is this one of our tanks – yes or no?', 'if no, is this a T72 – yes or no?', and, when combined, could lead to the AI application effectively outputting 'this is an enemy T72 tank'.

In some cases, the number of categories could be very large, and the AI has to allocate something to one of many categories directly, rather than via a series of binary classifications. An example of this is outputting exactly which ship has been sighted using a catalogue of the world's ships. It is a matter of the specific technique being used, the training data available and the skills of the development team that decides whether a cascaded or direct approach is best.

Regression, in the sense of AI and ML, concerns the prediction or output of continuous parameters. Whereas classification AI might output whether something is a tank, whose it is and which type is it, a regression AI might output how fast the tank is going or which direction it is about to turn. Both speed and direction have no natural categories although they do have boundaries: direction measured with polar coordinates and can be reported between 0 and 360 degrees, and speed can be reported between 0 and the maximum speed of the vehicle. Fitting a trendline, which can be done easily using Microsoft Excel, is an example of something that could form the basis of a regression algorithm. The same applies to the linear regression function within Excel and other applications such as 'R'. However, in practice, issues of precision and the number of significant figures and decimal places used in the output can turn regression outputs into classification problems.

There is a habit amongst data scientists and AI developers to allow their algorithms to output to the greatest precision possible, in other words with the most significant figures and decimal places. This can provide a false sense of accuracy but may not provide much practical benefit. For instance, if an AI application is used to predict the change of direction of an enemy armoured vehicle in order to better position a top-attack drone, there may be little difference between a prediction that it will turn to 32 degrees from north and a prediction that it will turn to 32.3504671 degrees from north.

This opens up the option that regression problems can, if fact, be turned into classification problems by dividing the continuous output space (such as speed or direction) into bands or bins that reflect different classes. In the example above, creating 1 degree classes could reduce the problem from a continuous 0 to 360 degrees to 360 bands, each 1 degree wide. This is still a lot, but if a precision of 10 degrees was determined to be acceptable, then the problem could be reduced to 36 bands of 10 degrees' width. In some cases it might be acceptable just to indicate whether the threat is moving north, south, east or west, so the problem has only four classes. The reason for turning regression problems with continuous outputs into classification problems with categorical outputs is that it increases the number of algorithms that can be applied, and thus makes it much easier to convert the raw outputs of AI algorithms into actionable outputs.

Converting classification problems into regression problems occurs less often but also has some benefits, as long as the classes can represent the progression from good to bad or low to high. As an example, for an AI algorithm that predicts whether an entity (such as an enemy aircraft or vehicle) is going to fire, the classes might be defined as 'yes', 'no' and 'maybe'. This classification could be transformed into a regression problem where the output is likelihood of firing,[4] which could take a value between 1 and 100. For decision making, all the user or operator needs to know is whether the answer is yes, no or maybe, and these categories could be overlaid onto the scale with user input and guidance. For example, the overlay might be that 1 to 35 is no, 36 to 65 is maybe and 66 to 100 is yes. The operator now has some level of implicit confidence in the output that allows adjusting their response depending on the situation. In peacetime, an output of 40 would count as a low maybe and might trigger a relaxed response, whereas in wartime an output of 60 would be a high maybe and may trigger a determined search for additional data to refine the output or protection measures, just in case the threat suddenly changes behaviour. Military operators have found this use of regression outputs overlaid with categorical outputs very useful, so use them often.

Baselining, Benchmarking and Data Balancing

With classification problems there are a wide range of measures and metrics that can be used and no doubt anyone who develops AI algorithms would love to explain why their new one is better than the older ones. Whilst new and (usually) complex metrics are effective for developing and tuning the AI, the metrics that are key to building evidence and trust with potential users are those that are easy for them to understand. Accuracy, false positive rate (FPR) and false negative rate (FNR) are three such easy-to-understand metrics.

Accuracy is usually expressed as the percentage of times the AI picks the correct classification. It can be easily calculated for a binary case or when there are many potential categories to choose from. In an ideal world accuracy would be 100 percent, but in reality it is likely to be far lower. That leaves the

---

4 Note we are referring to 'likelihood' rather than 'probability'. The difference is very important, particularly when dealing with predictive AI, but it is up to you as to whether you wish to do a search for the difference between probability and likelihood.

developers and users to work out how much accuracy is good enough and how to make use of the AI when it has a known level of error.

In a simple example of an AI application to detect whether a long-range radar track belongs to military aircraft, it could be trained using 1,000 labelled radar tracks – in other words, supervised learning – and then tested with 100 different radar tracks. Its accuracy would be the number of times that it correctly classified the 100 test tracks, for example 60 percent. The first question to ask is whether this is this considered a good level. This is where the baselines and benchmarks come in. The baseline represents the current level of performance, for instance with the existing non-AI systems or human capability. However, for a totally new capability a baseline would not exist. The benchmarks are standard rules, heuristics or algorithms that represent basic non-AI methods of outputting a classification, which complement the baseline, if one exists, or can provide an alternative basis for assessing the accuracy.

The two classic benchmarks are luck and Zero R. Luck represents the statistical outcome of random choices, such as flipping a coin (in the case of a binary choice as in this example). Over a large number of tests, this reduces simply to one over the number of options; in this case that would lead to an average accuracy of 50 percent, so the 60 percent achieved by the AI looks to be an improvement. The Zero R benchmark uses the details of the training data. For this example, 700 of the training data of 1,000 radar tracks were of military aircraft and 300 were of non-military aircraft. As the frequency of military aircraft is higher, the Zero R method would always output military aircraft (irrespective of the input data) as that would be correct 70 percent of the time, based on the training data. Using the test data, the accuracy of Zero R depends on the split of positives (military) to negative (non-military) cases. So, if 100 test data sets contained 70 military aircraft and 30 non-military aircraft, then Zero R would give an accuracy of 70 percent, which seems better than the AI. However, if the test data were split the other way, Zero R would give an accuracy of 30 percent, so much worse than the AI. This ability to make the AI look better or worse compared to benchmarks through splitting the training data versus splitting the test data highlights the importance of understanding the balance of the training and test data and the subsequent choice of the appropriate benchmarks (although the example above was chosen to highlight this issue – it does not always occur).

Test data sets should, as far as possible, represent a real-life split so that the AI is provided with a realistic test of its abilities. Real life can vary in different situations – in this case, peacetime would see many more civilian tracks on the radar and in wartime perhaps many more military tracks. This just means multiple tests with multiple data sets should be conducted in order to understand how well the AI performs in different situations. When training, however, it can sometimes be necessary to alter the normal split so that the AI has a similar number of learning datapoints for each potential output. This is often done by oversampling data that represents the less frequent outcome, and undersampling data that represents the most frequent outcome. Over- and undersampling must be limited to the training data set and not skew the test data set.

Where the test data set is naturally unbalanced, the raw accuracy value on its own can be misleading and that is when the Zero R benchmark becomes important. To continue with the example, both the training and test data had a 70/30 split between military tracks and non-military radar tracks (which is unbalanced). In a real-life profile, the 60 percent accuracy of the AI would not be considered good when the Zero R benchmark gives an accuracy of 70 percent.

However, additional insights from the FNRs and FPRs can be used to determine whether the AI could still be useful, but only in the binary classification case. The FNR is defined as the number of positive events wrongly categorised as negative (false negatives) divided by the total number of actual positive events. Similarly, the FPR is defined as the number of negative events wrongly categorised as positive (false positives) divided by the total number of actual negative events.

For the example in this chapter, there is enough information to calculate the FNR and FPR for the Zero R benchmark with the test data. Because applying Zero R to the training data (in which 70 percent of the data sets are military) leads it to assume all tracks are military in the test data, the number of false negatives (military tracks identified as non-military) is 0. Hence the FNR for the Zero R is 0 percent, which might seem desirable. However, the number of false positives given by Zero R in the test is 30 (the number of civilian radar tracks it incorrectly labelled as military) whilst the number of true positives (the number of civilian tracks in the test data set) is also 30. Hence the FPR for the Zero R benchmark is 30 divided by 30, which gives a ratio of one (in other words, 100 percent of the positives were false) which clearly is a bit worrying and highlights that the initial view of 70 percent accuracy was misleading on its own.

To calculate the FNR and FPR for the AI, more information is needed about how the AI performed. The AI's overall accuracy of 60 percent based on the test data from 70 military and 30 civilian radar tracks needs to be broken down further. For this example the AI identifies 45 of the 70 military tracks correctly as military but incorrectly identifies 25 as civilian. Of the 30 civilian tracks, it correctly identifies 15 as civilian but incorrectly identifies the other 15 as military. The 45 correctly identified military tracks and 15 correctly identified civilian tracks lead to the overall accuracy of 60 percent. The 25 out of 70 military tracks incorrectly identified as civilian tracks (false negatives in the sense of falsely being identified as not military) means the FNR is 36 percent. The 15 out of 30 civilian tracks incorrectly identified as military tracks (false positives in the sense of falsely being identified as military) means the FPR is 50 percent.

So far, the Zero R benchmark has come out top in terms of overall accuracy (70 percent versus 60 percent) and FNR (0 percent versus 36 percent), whereas the AI has come out top in terms of FPR (50 percent versus 100 percent for the Zero R benchmark). The final step is to consider what the system (including the human operator) is going to do with the information. If the AI were to be linked directly to an automated self-defence system that would shoot down all military tracks, then this AI and the Zero R benchmark would be unacceptable because 50 percent to 100 percent of civilian aircraft seen would be shot down, respectively. However, if this AI were to be used to prioritise radar tracks for the human team, then it has reduced the operator workload by 40 percent (the 25 military tracks incorrectly identify as civilian, and the 15 civilian tracks correctly identified as civilian). Of course, the AI could be reapplied as the radar inputs are updated, and the operators would eventually get down to the 40 percent at the bottom of the list. So, although the AI clearly needs to be improved, it has shown some potential benefit. These numbers were chosen to illustrate issues regarding metrics and the balance of data sets and should not be taken as representative of AI performance.

The example above was for classification AI. Now assume the AI is intended to predict the change of course of the military radar tracks, in degrees, at 5 second intervals. Accuracy in this case would not be in terms of the percentage of times the AI was correct but rather the average error in the prediction, that is the difference between the prediction and what actually happens. Care has to be taken in not allowing underestimates to cancel out overestimates, so often the average error is calculated purely in terms of magnitude or a root-mean-squared (RMS) value is calculated. It is

Timestep	Actual change in direction (degrees)	AI predicted change in direction (degrees)	Error (degrees)	Magnitude of error (degrees)
1	0	0	0	0
2	0	1	1	1
3	−4	−1	3	3
4	−2	1	3	3
5	0	−1	−1	1
6	0	0	0	0
7	3	−1	−4	4
8	6	4	−2	2
9	2	0	−2	2
10	0	0	0	0

Table 8.1: Example data for an AI application that predicts a threat's change in direction

also important to take into account the spread of the values by looking at the maximum and minimum errors as well as the standard deviation in errors.

As an example, Table 8.1 shows the actual and predicted changes of direction for ten timesteps, along with the calculation of the error (taking into account the direction of the error, where a negative indicates a low prediction and a positive indicates a high prediction), and the magnitude of the error.

The average error in the AI prediction is only −0.2 degrees, which is deceptively small because it is the result of the positive and negative errors cancelling each other out. However, the spread of the error is from −4 degrees (4 degrees too low) to 3 degrees (3 degrees too high). Calculating the average from the magnitude of the errors is more useful and indicates that the AI predicts the change in direction with an average error of 1.6 degrees.

How does this analysis compare to potential benchmarks? Equivalent benchmarks to luck and Zero R can be applied here. Luck would be to assume the average change in direction observed always applies (in this case, 0.5 degrees). In this example, the spread of error would be from 5.5 degrees too low and 4.5 degrees too high, with an average magnitude of error of 1.8 degrees, which is clearly worse than the AI on all levels. The Zero R benchmark would be to assume the most frequent change, which is actually 0 degrees (meaning most of the time it does not change direction). Here the spread of error would be from 6 degrees too low and 4 degrees too high, with an average magnitude of error of 1.7 degrees, which is still worse than the AI on all levels. The key desire for regression AI is that the spread of error is as narrow as possible, as this can then be corrected for.

This analysis could be converted to a classification problem by defining five classes: where the change in direction is less than 1 degree either way, where the change is between 1 and 3 degrees to port, where the change is more then 3 degrees to port, where the change is 1 and 3 degrees to starboard, and where the change is more then 3 degrees to starboard. If this were done the AI would have an accuracy of 60 percent. Whether this is useful depends, again, on how the outputs of the AI are to be used. If the AI is intended to help develop a fire-control solution, then the precision of the regression output may be beneficial. If it is to make a decision on courses of action or threat intent, the conversion to a classification problem may be more useful.

Hierarchy of Metrics

So far the measures discussed are at the level of the AI performance. These represent the lowest level of performance of the overall socio-technical system, and largely fall under the purview of AI developers. For military sponsors and those responsible for requirements, scrutiny and procurement, these metrics need to fit into the operational context and contribute to operational outcomes.

The Code of Best Practice (CoBP) for C2 Assessment[5] set out by NATO for assessing command and control (C2) performance has proved popular in a range of AI projects. This is because AI, whether for robotic and autonomous systems or digital systems, is a form of C2, albeit within a wider system. The key aspect of the CoBP is the hierarchy of measures it defines that links the lowest measures of performance (MoPs) at which the AI sits to the highest level of measures of outcome (MoOs). These levels have been refined over the years and the definitions used across a wide range of experimental and procurement projects for both C2 and AI capability in the last few years are as follows.

At the highest level sit the MoOs, also referred to as measure of force effectiveness (MoFE). These are usually applied to a military unit or

---

5 Command and Control Research Programme, *Code of Best Practice for C2 Assessment*, (NATO, 2002).

force (such as a naval task force) and indicate whether a success or failure was achieved – for example, mission success. These can be binary (mission achieved or not) or could have some index of success such as number of ships in the naval task force that were put out of action.

Next come the measures of effectiveness (MoEs). These are applied to the defence capability that the system is delivering, and measure the capability's contribution to outcome. This capability is typically a socio-technical system that combines technical systems with human teams and processes. An example is the maritime air-defence capability on a naval destroyer which encompasses the operations room's combat management system (CMS), the team of operators in the ops room, the missiles used to shoot down a threat (referred to as hard kill), various soft-kill mechanisms such as chaff and decoys, and close-in self-defence weapons such as Phalanx and Goalkeeper. In the case of this maritime air-defence example, suitable MoEs could be the number of threat missiles shot down, how close the threat missiles got, how many of the hard-kill and soft-kill systems are left, how many civilian aircraft were shot down. This is the combined effect of all the elements – the ops room and the hard-kill and soft-kill systems.

Next come the measures of system performance (MoSPs), which are applied to a specific system within a capability and measure its ability to fulfil its particular function. This might still be a combination of technical and human elements that make up a system. In the maritime air-defence example, the ops room (consisting of the team, the processes and the CMS) would be a system against which MoSPs could be gathered. Examples include the time taken to reach a decision on the threat, the time taken to give the order to act and the correctness of that action.

MoPs are individual parameters that indicate a component of the system's ability to contribute to a process. This is where the human and the technical parts of the system are separated, but both could be measured. For instance, with decision aides that could be used to assist in maritime air-defence (one of the areas AI has been tested) the human element of the system could be assessed using metrics such as workload and situational awareness, whilst the AI component of the system might have MoPs such as accuracy, false positive and negative rates, level of trust achieved (trust in the AI and how well it can be teamed with, now measured in experiments fairly regularly using several established indexes and measurement approaches).

Clearly, AI developers and those at the research, concept and requirement end of the procurement process will focus initially on the AI

MoPs (and indeed most of this chapter has done just that). Hopefully, it is evident that by linking these low-level metrics to the combined human and technical metrics (MoSPs), then to the capability level (MoEs) and finally to things that matter to the operational commander (MoOs and MoFEs), it becomes possible to understand the overall impact of different AI tools. This benefit chain can make rapid development, procurement and introduction so much easier if it is thought of up front rather than left to just the formal procurement process or – even worse – when government scrutineers demand evidence of the cost-benefit of the AI under consideration compared to any number of other ways government can spend the money.

9

Conclusions

Hearing is not as good as seeing, seeing is not as good as knowing,
knowing is not as good as acting;
true learning continues until it is put into action.
— Xunzi[1]

This book began with a caution about the hype surrounding artificial intelligence (AI) in defence – about the devotees who promise too much and the doomsayers who warn of catastrophe. Both extremes are often fuelled by people with little first-hand experience of developing, implementing or using AI. In contrast, this book has explored the practical experiences of four diverse projects that navigated the journey from concept to use. Together, they provide valuable insights for avoiding the cycle of overpromise, overspend and underdeliver by acting on the lessons learned, in the spirit of Xunzi's exhortation.

As the introduction noted, AI's potential may not lie in a single 'killer app' but in its ability to permeate defence in the same way that technologies such as the microchip or the internal combustion engine did – gradually, unevenly and in ways only fully understood through widespread adoption. To reach that point, defence organisations must avoid the 'trough of disillusionment' that follows inflated expectations. These four cases suggest that the way to do so is not to search endlessly for utility and then seek utilisation, but to recognise that utilisation itself drives utility.

This book ends with a summary of the common lessons and themes from these first-hand accounts of the experience of developing and implementing AI for defence use, albeit at small-scale and in niche areas. They are from a time when AI was relatively new and these cases were at the bleeding edge of applying AI to defence. It is impossible to predict how long these lessons will

---

1   Usually misattributed to Confucius, possibly because Xunzi was a follower of Confucius.

be relevant, as the technology and commercial landscape change – although perhaps there is some AI that could. So these are offered as reflections rather than predictions.

Ensuring Utilisation

The utilisation staircase consists of 10 questions whose answers were, in different ways, key to moving from concept to defence use:

1. What existing (human) process is it addressing or is it similar to?
2. What is the difference to other AI?
3. What does the AI do better, or allow one to do, that was not possible before?
4. Is it as complex as needed (simple as possible) or as complex as possible?
5. What perceived risks does it deal with or are not relevant?
6. Is there enough time to mature and integrate?
7. Do we really understand how the military user would use it?
8. What is the level of ease of use?
9. What risks does it introduce?
10. What will be done to address the new risks?

Although the specific answers varied by case, there were five lessons that were common to all and which helped achieve utilisation.

1. Start with the human process

Each successful AI project began by asking what existing human process it addressed. DUCHESS mimicked and scaled up interviews, MALFIE filtered and explained AI outputs to overwhelmed operators, Red's Shoes mirrored how commanders anticipate adversary learning, and DR SO compressed years of tactical learning into hours of simulation. In every case, AI was most valuable not as a replacement for human decision making, but as an augmentation of it.

2. Favour simplicity over complexity

Across the cases, the most effective AI was as complex as needed, and as simple as possible. DUCHESS succeeded because it delivered adaptive conversations without intimidating users. MALFIE thrived by translating technical alerts into human language. Even DR SO, based on deep reinforcement learning, distilled its outputs into doctrine-like 'playbooks' operators could understand. The temptation to adopt cutting-edge techniques was resisted in favour of functionality and trustworthiness.

3. Build trust iteratively

Trust was not assumed; it was earned. This required careful attention to what perceived risks the AI deals with, and which are not relevant. DUCHESS overcame fears about anonymity and bias. MALFIE alleviated concerns about information overload. Red's Shoes, working at the operational level, struggled longer because its developers lacked an intuitive understanding of high level headquarters culture. DR SO benefitted from this lesson by involving users early and often, even inviting them to critique errors in the AI's outputs.

4. Cultural alignment is decisive

In all four cases, cultural understanding between AI developers and military users was the most critical – and often the most fragile – factor in success. Where this alignment existed, as in DUCHESS and MALFIE, adoption was rapid. Where it was lacking, as in Red's Shoes, years were added to the journey. DR SO showed how cultural alignment can be built deliberately through iteration and co-creation.

5. Participation accelerates progress

Finally, the cases illustrate the importance of participation. AI developers who stepped into the military users' shoes and military officers who engaged directly with the AI created mutual understanding and accelerated adoption. This participatory mindset contrasts with the passive model of defence acquisition that too often sees AI as 'just another technology' to be delivered by contractors. Having those contractors working within the MOD building or as part of an MOD team does not, in itself, solve the problem. It takes a team where the contractors are able to understand the military user and use case well enough to be able to act as the user, and the military needs to know the AI well enough to know when it is useful and when it is not.

Diversity

Just as humans are extremely diverse in every way (including in their level and type of intelligence), so too is AI. Given that governments and defence organisations have defined AI as technology that allows the use of a computer to make decisions normally made by humans, the term can cover everything from computationally simple rulesets to hugely complex neural nets and large language models. The people and organisations that develop AI algorithms and applications range from huge global companies such as Apple, Microsoft and Palantir down to micro businesses developing the AI described in this book. The application user base can be in the hundreds of millions, as for ChatGPT, down to one for niche use cases. Use cases can range from suggesting tasks for a human operator working in a back-office function through to autonomously controlling systems and weapons. So the phrase 'AI will...' should always prompt questions about which type of AI, produced by whom and used in what way, in order to understand its role.

The driving force behind a firm's development of AI is another point of difference. With the amount of money flooding into AI all over the world, most AI firms have a particular technology or application that they are selling. Some, however, turned to AI to address a specific problem and then sought to commercialise their application further. Neither approach is necessarily good or bad, but each leads to different development and growth paths. Starting with the problem and finding the right AI can produce something that can be used quickly, as was the case with MALFIE. However, it is often then difficult to scale up and commercialise. Also, it needs regular updates as technology advances, as was found with DUCHESS. Starting with the AI is more of a risk but better for scaling and commercialisation, but it places the onus on the (potential) user to make the best of it.

Moreover, over time, the diversity of expertise within the firms developing AI for defence (which was a feature of these cases) may fade. So many companies are developing in-house AI capabilities that the only unique selling point will be a high level of knowledge in a narrow area. But if a key enabling factor, such as access to data or computing power (or energy), becomes concentrated in a few companies' hands, things might move the other way, with a few firms doing lots of different AI things.

Assumptions

With the diversity in AI comes a diversity in the relevant assumptions for each AI project. Typical assumptions are that a lot of data is needed, that it is only 'true AI' if it continues to learn as it is being used, that a person cannot understand or explain what the AI is doing, and that there are ethical issues to be investigated. All of these can be true, but in many cases they need not be. Alas, there is a financial imperative for AI firms to paint every AI use case and opportunity as having all these issues. Anyone involved in developing, procuring or implementing AI in defence needs to be alive to the possibility that there may be simpler and less risky options in terms of AI techniques and approaches than those being offered.

Only half the cases in this book needed to gather lots of training data (DUCHESS and MALFIE) and in neither case was it a practical problem. Red's Shoes was specifically developed to work with small amounts of data, and DR SO generates its own as part of its reinforcement learning. The focus on big data in many AI projects is partly because a lot of AI capability came out of data science work. Another possibility is that there is human bias that more information must lead to a better decision, combined with the availability of computing power to process the data. Yet, just as natural intelligence has the ability to make decisions on minimal information and computing power, so too are there plenty of techniques and use cases where the AI can (and perhaps should) work with little data.

In terms of the need and risk of the AI learning as it goes, in all the cases in this book it was found that it was desirable and practical to separate the process of the AI's learning from the process of the operators' usage of the AI application. Just as individuals in the military undergo refresher training and regular testing, the AI can undergo further training and updates, just on a quicker cycle than humans do. Again, there has been a commercial incentive to paint only certain techniques as 'true' AI, but once enough organisations have an AI capability they will see that there are plenty of less fancy techniques that provide AI functionality at lower cost and risk.

Three quarters of the cases (MALFIE, Red's Shoes and DR SO) involved incorporating functions that allowed the operator to understand the AI's outputs. This is a developing field with many different approaches and opinions, so best practices have not yet been shared. However, by taking a fairly standard 'design-thinking' approach combined with a proactive user-feedback process it became possible to get military operators engaging with the AI in ways similar to how they would engage with a new and slightly odd member of their team.

Alongside assumptions about risks, it was found again and again that assumptions about language can be tricky. Arguments over the definition of AI caused many issues in the early stages, even after the Ministry of Defence (MOD) came up with its definitions. Thankfully, that seems to have passed. However, definitions still proved a problem at the application level. Referring to DUCHESS as an 'AI interviewer' often led to people assuming it was for job interviews. In MALFIE the word 'anomaly' was used when 'pattern of interest' was meant. For DR SO, the word 'swarming' caused confusion. In all cases much time and effort were spent defining what was meant.

Culture

With the benefit of hindsight, the biggest driver and barrier for getting these AI applications into use was the level of cultural understanding between the AI development team and the military user. It may have been a matter of luck that groups of ex-MOD or ex-military had been able to bring AI techniques in to help address problems that were known and understood in their military context.

DUCHESS and MALFIE were, in this respect, straightforward. Conversational interviews for feedback, lessons learned and investigations are all done in civilian as well as military life. For the initial DUCHESS proof of concept, it was necessary to make allowances for 'Jack Speak' (words, language and euphemisms peculiar to the RN), but there were documents that could be used to train the AI to recognise and translate these. MALFIE required a little more engagement with naval officers to understand how they might use something that prioritises potential anomalies. However, it was easy to draw good parallels to similar challenges in civilian life.

Red's Shoes was a much more challenging case. Although many of those involved had experience of military decision making, it was largely at the tactical level. Operational decision making was, qualitatively, very different. The developers had read the relevant doctrine and so knew, theoretically, what decisions the Red's Shoes AI could contribute to. But they lacked an understanding of the culture and process of a higher-level headquarters and, hence, the awareness of when, where and how the AI's inputs and outputs would be managed. This showed in the length of time it took to go from proof of concept to an actual military user, which reflected the length of time needed to align themselves to the culture (and organisation) of operational decision making.

DR SO would have been more challenging had these lessons not been learned. Creating opportunities to iterate the AI outputs with the military was built into the plan right from the start. Military stakeholders' views were actively sought on what the AI had gotten wrong, rather than looking for what they liked. This improved their cultural awareness quicker than would have been possible otherwise. Once the team thought they understood enough about the culture and process of the current military process, they took upon themselves the task of writing the concept of operation of the AI. Again, this was as much to get the military stakeholders to point out where things might be wrong. By doing this, the team fully filled the gap between military culture and understanding and AI development community culture and understanding.

Participation

Since World War Two, the western attitude towards defence has been that it is the domain of a relatively small group of professionals. The Falklands, First Gulf War and Balkan wars provided some jolts to this comfortable view in the UK. The 9/11 attacks and subsequent operations in Afghanistan and Iraq raised awareness, but mostly of the human cost. Russia's invasion of Ukraine in 2022 shook the foundations. One of the things that is most telling about Ukraine is the extent to which people have found ways to participate. The large number of drone initiatives and cyber units set up by civilian organisations are two examples. These mirror many significant changes in military technology where it was participation rather than performance that mattered. For instance, tests by Sir Ralph Payne-Gallwey[2] demonstrated that the bow and arrow remained more effective than firearms in terms of range and accuracy until the mid-19th century. However, it was much easier to train a musketeer than an archer, hence the switch. Now, it is much easier to be a drone operator than a fighter pilot or sniper, which makes it easier for people to participate in the defence of their country.

---

2 Ralph Payne-Gallwey, *The Book of the Crossbow* (Dover Books, 1907).

A key part of the success of these four cases was encouraging the AI developers to put themselves in the military user's shoes and participate in the decisions and processes they were trying to enable or improve with the AI. Similarly, the military officers involved with these projects were encouraged to 'give the AI a go' as often as possible. However, this was during a period when access to AI was limited to specialists with access to suitable computers and facilities.

Now, using technology such as ChatGPT, it is possible to get AI to write algorithms for anyone to use, with little prior AI knowledge. So anyone who has to take responsibility for any aspect of AI should experiment with developing their own basic AI. They are unlikely to develop something that would outperform what a professional AI company might offer, but they will gain a better understanding of what they are being told and offered. Through better understanding, anyone can make a greater contribution to the level and speed of impact that AI can have on defence capability.

BIBLIOGRAPHY

Allen, G. *Understanding AI Technology* (Joint AI Centre, US DOD, April 2020)

Arens, M. "Lessons Learned from Hezbollah", *Haaretz*, last modified 23 October 2017, https://www.haaretz.com/opinion/. premium-lessons-learned-from-hezbollah-1.5447889.

Arquilla, J. and Ronfeldt, D. *Swarming and the Future of Conflict* (RAND, September 2000).

Bechara, A., Damasio, H. and Damasio, A. "Role of the Amygdala in Decision-Making", *New York Academy of Sciences*, Vol 985, Issue 1 (The Amygdala in Brain Function: Basic and Clinical Approaches, pages 356–369) April 2003, https://pubmed.ncbi.nlm.nih.gov/12724171/.

Biddle, S. *Military Power – Explaining Victory and Defeat in Modern Battle* (Princeton University Press, 2006).

Bradley, T. "Amazon Statistics: Key Numbers and Fun Facts", https://amzscout.net/blog/amazon-statistics/, accessed 27 December 2023.

Command and Control Research Programme, Code of Best Preactice for C2 Assessment (NATO, 2002).

Counsel Direct, "A Brief History of Speed Limits" (8 February 2015). Accessed 31 March 2024. http://www.counsel.direct/news/2015/2/8/a-brief-history-of-speed-limits.

Dibbert, T. "How Sri Lanka Won the War", *The Diplomat*, 2 April 2015, https://thediplomat.com/2015/04/how-sri-lanka-won-the-war/.

Edwards, S. *Swarming and the Future of Warfare* (RAND, May 2005).

Edwards, S. *Swarming on the Battlefield: Past, Present and Future* (RAND, March 2000).

Everstine, B. "Integration of HARM on Ukraine's MiG-29s, Su-27s Took 2 Months," Aviation Week Network (20 September 2022). Accessed 25 November 2024. https://aviationweek.com/defense/missile-defense-weapons/integration-harm-ukraines-mig-29s-su-27s-took-2-months.

Falcon, A. "Aristotle on Causality," *Stanford Encyclopedia of Philosophy*, last modified 7 March 2023. Accessed 25 November 2024. https://plato.stanford.edu/entries/aristotle-causality/.

Fraassen, B. *The Scientific Image* (Oxford: Clarendon Press, 1980).

Gillespie, N., Lockey, S., Curtis, C., Pool, J. and Akbari, A. *Trust in Artificial Intelligence: A Global Study on the Shifting Public Perceptions of AI* (University of Queensland and KPMG Australia, 2023).

Gleis, J. "Ten Lessons Learned by Hezbollah", Huffington Post, last modified 2 February 2015, https://www.huffingtonpost.com/joshua-gleis/ten-lessons-learned-by-he_b_6564850.html.

GlobalSecurity.org, "Second Chechen War". Accessed 25 November 2024, https://www.globalsecurity.org/military/world/war/chechnya2.htm.

Gudmundsson, B. *Stormtroop Tactics: Innovation in the German Army, 1914–1918* (Praeger Paperback, 1995).

Hempel, C. *Aspects of Scientific Explanation and Other Essays in the Philosophy of Science* (New York: Free Press, 1965).

Higher Education Statistics Agency, "Table 50: Students by Subject Area and Sex". Accessed 31 March 2024. https://www.hesa.ac.uk/data-and-analysis/students/table-50.

Internet Encyclopedia of Philosophy, "Explanation". Accessed 25 November 2024. https://www.iep.utm.edu/explanat/.

Jaffri, A. "Explore Beyond GenAI on the 2024 Hype Cycle for Artificial Intelligence", last modified 11 November 2024, https://www.gartner.com/en/articles/hype-cycle-for-artificial-intelligence.

Jaya-Ratnam, D. *DRAM Busters: The Coming Slump* (FOR Securities, July 2003).

Kahneman, D. and Tversky, A. "Prospect Theory: An Analysis of Decision Under Risk", *Econometrica*, Volume 47, March 1979.

Klein, G., Orasanu, J., Calderwood, R. and Zsambok, C. *Decision Making in Action: Models and Methods* (Ablex Publishing, 1993).

KPMG, "Trust in Artificial Intelligence: A Global Study (2022)". Accessed 25 November 2024. https://ai.uq.edu.au/files/6161/Trust%20in%20AI%20Global%20Report_WEB.pdf.

Lagrone, S. Navy: Saudi Frigate Attacked by Unmanned Bomb Boat, Likely Iranian (USNI News, last updated 20 February 2017). https://news.usni.org/2017/02/20/navy-saudi-frigate-attacked-unmanned-bomb-boat-likely-iranian.

Ledwidge, F. *Losing Small Wars: British Military Failure in Iraq and Afghanistan* (Yale University Press, 2011).

Lukowiak, K. *A Soldier's Song* (Secker & Warburg, 1993).

Marsh, G. "Military Upgrades: How the Royal Navy Advanced Its AEW," Avionics International (1 July 2001). Accessed 25 November 2024. https://www.aviationtoday.com/2001/07/01/military-upgrades-how-the-royal-navy-advanced-its-aew/.

McNerney, M. et al. *National Will to Fight: Why Some States Keep Fighting and Others Don't* (RAND, RR2477, 2018).

Montier, J. *Seven Sins of Fund Management: A Behavioural Critique* (Dresdner Kleinwort Wasserstein, November 2005).

NATO Communications and Information Agency, "NATO's Joint Analysis and Lessons Learned Centre: On the Cutting Edge of Innovation," NATO, 3 May 2023. Accessed 2 August 2025. https://www.act.nato.int/article/natos-joint-analysis-and-lessons-learned-centre-on-the-cutting-edge-of-innovation/.

O'Connell, R. *Of Arms and Men: A History of War, Weapons and Aggression* (Oxford University Press, 1990).

Oxford Dictionaries, "Explanation". Accessed 17 August 2025. https://www.oed.com/dictionary/explanation_n?tab=factsheet#5067609.

Pallotta, G., Vespe, M. and Bryan, K. "Vessel Pattern Knowledge Discovery from AIS Data: A Framework for Anomaly Detection and Route Prediction", Entropy (2013). https://doi.org/10.3390/e15062218.

Payne-Gallwey, R. *The Book of the Crossbow* (Dover Books, 1907).

Perret, B. *At All Costs: Stories of Impossible Victories* (Cassell Military Classics, 1998).

Plous, S. *The Psychology of Judgement and Decision-Making* (McGraw-Hill, 1993).

Rothchild, J. "When the shoeshine boys talk stocks: It was a great sell signal in 1929. So what are the shoeshine boys talking about now?" (*Fortune*, 15 April 1996). Retrieved from https://money.cnn.com/magazines/fortune/fortune_archive/1996/04/15/211503/.

Scott, R. "Machine speed warfare: UK tests naval AI decision aids in ASD/FS-21 exercise", *Jane's International Defence Review* (July 2021).

Shishani, M. 'The Chechen Mujahideen and the War in Iraq', *Journal of Slavic Military Studies* Vol 18, No 4 (2005): doi:10.1080/13518040500355015.

Statista, "Trust in Doctors Worldwide by Country". Accessed 31 March 2024. https://www.statista.com/statistics/1274258/trust-in-doctors-worldwide-by-country/.

The Peak Performance Center, "Skill Will Matrix". Accessed 25 November 2024, http://thepeakperformancecenter.com/business/coaching/skill-will-matrix/.

Torres, M., Hart, G. and Emery, T. *The Dstl Biscuit Book: Artificial Intelligence, Data Science and (Mostly) Machine Learning* (DSTL/PUB115968, 2019).

UK Ministry of Defence "Assurance of Artificial Intelligence and Autonomous Systems: a Dstl biscuit book", last modified 1 December 2021, https://www.gov.uk/government/publications/assurance-of-ai-and-autonomous-systems-a-dstl-biscuit-book/assurance-of-artificial-intelligence-and-autonomous-systems-a-dstl-biscuit-book

UK Ministry of Defence, "Campaign Planning", Joint Doctrine Publication (JDP) 5-00", second edition, chapter 2, section 4, https://assets.publishing.service.gov.uk/government/uploads/system/uploads/attachment_data/file/434557/20150609-JDP_5_00_Ed_2_Ch_2_Archived.pdf.

UK Ministry of Defence, "Red Teaming Guide", Chapter 4 – 'Applying red teaming to defence problems', https://assets.publishing.service.gov.uk/government/uploads/system/uploads/attachment_data/file/142533/20130301_red_teaming_ed2.pdf.

UK Ministry of Defence, Ambitious, Safe, Responsible: Our Approach to the Delivery of AI-Enabled Capability in Defence – Annex A: Ethical Principles for AI in Defence (UK MOD, 15 June 2022).

UK Ministry of Defence, Ambitious, Safe, Responsible: Our Approach to the Delivery of AI-Enabled Capability in Defence – Annex B: The Ministry of Defence AI Ethics Advisory Panel (UK MOD, 15 June 2022).

UK Ministry of Defence, Defence Artificial Intelligence Strategy (UK MOD, 15 June 2022).

UK Ministry of Defence, JSP 936 Dependable Artificial Intelligence (AI) in Defence (UK MOD, February 2024).

UK Ministry of Defence, JSP 936 Dependable Artificial Intelligence (AI) in Defence, February 2024.

University College London, Honours Degree Outcomes Statement 2019/20 (UCL, 2020).

Unknown, "The Other Side of the Coin: The Russians in Chechnya", *Small Wars Journal*, accessed 25 November 2024, https://smallwarsjournal.com/jrnl/art/the-other-side-of-the-coin-the-russians-in-chechnya.

US Department of Defense, Summary of the 2018 Department of Defense Artificial Intelligence Strategy –Harnessing AI to Advance Our Security and Prosperity (US DoD, February 2019).

Watling, J. and Reynolds, N. *Meatgrinder: Russian Tactics in the Second Year of Its Invasion of Ukraine* (Royal United Services Institute, May 2023).

INDEX

www.ingramcontent.com/pod-product-compliance
Ingram Content Group UK Ltd.
Pitfield, Milton Keynes, MK11 3LW, UK
UKHW020105121025
463846UK00004B/31

9 781912 440740